AMERICAN HIGHER EDUCATION, LEADERSHIP, AND POLICY

American Higher Education, Leadership, and Policy

Critical Issues and the Public Good

Penny A. Pasque

Foreword by Edward P. St. John and Lesley A. Rex

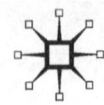

AMERICAN HIGHER EDUCATION, LEADERSHIP, AND POLICY

First published in hardcover in 2010 by PALGRAVE MACMILLAN® in the United States—a division of St. Martin's Press LLC, 175 Fifth Avenue, New York, NY 10010.

Where this book is distributed in the UK, Europe and the rest of the world, this is by Palgrave Macmillan, a division of Macmillan Publishers Limited, registered in England, company number 785998, of Houndmills, Basingstoke, Hampshire RG21 6XS.

Palgrave Macmillan is the global academic imprint of the above companies and has companies and representatives throughout the world.

Palgrave® and Macmillan® are registered trademarks in the United States, the United Kingdom, Europe and other countries.

ISBN: 978–1–137–45445–4

Library of Congress Cataloging-in-Publication Data is available from the Library of Congress.

A catalogue record of the book is available from the British Library.

Design by Newgen Knowledge Works (P) Ltd., Chennai, India.

First PALGRAVE MACMILLAN paperback edition: November 2014

10 9 8 7 6 5 4 3 2 1

To the next generation of college students including Cassie, Maggie, Brooke, Leo, Annabella, Lively, and baby Natalie.

Contents

FIGURES AND TABLES

FIGURES

TABLES

Foreword

It is rare that a recent PhD is able to write a book that has something profound to say to the higher education community and its leadership, but in *American Higher Education, Leadership and Policy: Critical Issues and the Public Good*, Penny Pasque exceeds this high standard of contribution. Dr. Pasque has written a book that confronts and critiques the tacit assumptions of educational leaders who dominate the conversations that occur behind closed doors.

The United States is suddenly more diverse than before and, finally, has a president who represents this diversity. President Obama shows the potential for bridging the great gap between discourses controlled by the powerful and the voices that emerge from the experiences of people within the communities, schools, and colleges that make up this diverse democracy.

In her research, Dr. Pasque had access to the private conversations of high-level leaders as they discussed their images of the public good in relation to the current trajectory of higher education in this nation. Clashing images of the public good dominated these conversations, just as they have dominated the literature and public policy over the past thirty years. At best, the espoused notions that divide the public discourse are feeble attempts to reconcile the neoliberal rationales of individual gain with the need for broader access and increased funding. What becomes clear in Dr. Pasque's text is that voices of women and people of color have been systematically quieted in debates on the definitions of public good and strategies to achieve it.

The clash between the two camps of conventional values becomes evident in this outstanding book. The older notions echo through the halls of Congress and dominate the air waves, but these loud voices are no more important than deeply held concerns about inequality among diverse citizens who have been silenced for so long.

It takes powerful research to break through entrenched notions and systems of ideas and values about public good and who can speak for it. The investigation that undergirds Dr. Pasque's arguments meets that criterion. Her study takes its power from established social science methods, while trailblazing a novel approach to designing and conducting research in higher education. She focuses on discourse—local dialogues and the societal values and beliefs that animate them. She fine-tunes her conceptual lenses to bring into view race, ethnicity, gender, sexual orientation, and class in the discourse practices of heavily invested higher education gatekeepers. And, as though stopping time and holding their discourses under a microscope, she dissects their consequential meanings, laying open a view that the public has rarely been able to access, let alone scrutinize.

Through Critical Discourse Analysis, the study peers into the social identities performed and created and the values enacted and reified in the dialogues. We see not only that talk matters but also *how* it matters and to whom. By transcribing the talk that so often flies unexamined across meeting rooms, the renderings supplant the domineering influences of time and space. Moments become permanently fixed so that analytical interpretations can reveal why they matter. Dr. Pasque illuminates why talk matters during moments of reflection on what is meant by *education for the public good*. More importantly, the remarkable strength of her analysis is in the way she connects dozens, scores, hundreds of moments to illustrate how during important dialogues people and their ideas are positioned into and out of power. We are shown how having one's say is a fraught and consequential event for those with less institutional authority and their important agendas.

So how can discourse analysis, coupled with open critical reflection on the definition of the public good, help us to overcome the conflicted condition of higher education policy in the United States? As Dr. Pasque's book eloquently argues, the voices of those who have been left out must be heard, as they are the basis for informing the redefinition of the public good and how we might achieve it. It is no longer tolerable, nor economically wise, to leave people out who don't fit molds of the wealthy and powerful. The strategies used to fund and guide higher education for three decades—privatizing public colleges by emphasizing benefits to those who can pay or borrow—must be reconstructed, just as educational leaders and policymakers must take time to listen to others in the room.

We may have a new president who inspires many of us to take fresh stances on critical issues, but we still need to listen more before we throw these ideas into the discursive space of public and private meetings that shape public policy. Indeed, the president himself runs into clashes between neoliberals and neoconservatives—the dialectic of the powerful—as he attempts to create new discourses on the future of the economy, health care, and education. As the shouting out at public forums on health care so boldly demonstrated, it is not easy to broaden the conversation, to get past old and worn out images of what works and what should not be tried. It is time for fresh thinking. Dr. Pasque's concepts of discourse illuminate possible new approaches to studying tactics used by those who attempt to change the policy conversation and public policy.

Edward P. St. John
Algo D. Henderson Collegiate
Professor of Education
Center for the Study of
Higher and Postsecondary
Education
University of Michigan

Lesley A. Rex
Professor Educational
Studies
Co-Chair of the Joint
Program of English
and Education
University of Michigan

Acknowledgments

I am extremely thankful to the people who participated in this research study and helped with the writing of this book. In particular, I deeply appreciate the sponsoring and co-sponsoring organizations who took the initiative to gather higher education leaders (university presidents, legislators, faculty, administrators, community partners, graduate students, foundation officers, national association administrators, etc.) to discuss the relationships between higher education and society. This group of leaders was willing to open themselves up to unintended scrutiny in order to help higher education leaders continue to learn from themselves/ourselves. To me, this effort takes a solid step toward connecting knowledge and action in the hopes of working toward educational and organizational change.

In addition, thank you to Drs. Lesley Rex, Ed St. John, John Burkhardt, Mark Chesler, and Pat Gurin for your mentorship and generous support. I have appreciated your insight and guidance over the years. I also want to thank my "disruptive dialogue" colleagues, Rozana Carducci, E. Ryan Gildersleeve, and Aaron Kuntz, for consistently challenging me to take the work to a deeper level. I also appreciate Rozana Carducci and Nick Bowman for their willingness to review drafts and help me sharpen my thinking and writing.

I would also like to thank the University of Oklahoma and the Research Council's Junior Faculty Program, for funding my efforts on this project with a grant during the summer of 2009. In addition, I am grateful for the University of Oklahoma College of Education's grant support the previous summer to work, in part, on chapter 6 of this volume, "Behind Closed Doors: Women's Voices of Resistance," which was presented at Oxford University, Oxford England, in August 2008 and is included here in its revised form. Bill Moakley, Director of Communications in the College of Education and Pam Harjo, Graduate Assistant, thank you both for help with the figures and tables in this book. I could not have produced "camera ready" materials without the two of you.

Finally, I owe a debt of gratitude to my two editors at Palgrave Macmillan, Burke Gerstenschlager and Julia Cohen, and to their amazing staff, including editorial assistant Samantha Hasey. I sincerely appreciate your effort and support with the publication process.

CHAPTER 1

Introduction to the Contemporary Context

The relationships between higher education and society are changing in the twenty-first century. Changes are taking place in terms of who pays for college, who gains access to college, and the universities' role in the global marketplace. For example, there have been decreases in public support for higher education (KRC Consulting, 2002; McMahon, 2009; Porter, 2002) and in state funding for public colleges and universities (Brandl & Holdsworth, 2003; Cage, 1991; Hansen, 2004), at a time when state and federal policies have linked higher education to the market in order to create jobs and increase economic viability (Bok, 2003; Jafee, 2000; Slaughter & Rhoades, 1996, 2004).

Recent national and global economic changes have caused ripple effects beyond Wall Street and Main Street; the ramifications have reached what I term Martin Luther King Jr. Boulevards across urban areas and College Avenues from coast to coast. Paul Krugman (2009), recipient of the 2008 Nobel Prize in economics, characterizes the situation this way:

> I'm tempted to say that the crisis is like nothing we've ever seen before. But it might be more accurate to say that it's like everything we've seen before, all at once: a bursting real estate bubble comparable to what happened in Japan at the end of the 1980s; a wave of bank runs comparable to those in the early 1930s (albeit mainly involving the shadow banking system rather than conventional banks); a liquidity trap in the United States, again reminiscent of Japan; and, most recently, a disruption of international capital flows and a wave of currency crises all too reminiscent of what happened to Asia in the late 1990s. (p. 165–166)

These "all at once" effects on Martin Luther King Jr. Boulevard and College Avenue are less of a focus in the mainstream media, but the crisis has nonetheless impacted the daily lives of people across the United States. The economic issues are forcing many students, potential students, and parents to weight their academic options in ways like never before as articles with titles such as "Why Don't Colleges Cut Costs, Tuition?" (Erb, 2009) and "What Is a Masters Degree Worth?" (Taylor et al., 2009) flood local newspapers across the country.

Although some crises have improved since 2008 and 2009, the ramifications of the economic downturn on College Avenue remain and include the reduction of endowments, furloughs, the rising costs of college, students' ability to pay, cancelation of student-centered co-curricular programs, and the struggle for survival of the local college town gift shop, to name a few. This shift, however, began prior to the recent economic changes and is reflected in an increase in the commercialization of higher education and academic capitalism (Bok, 2003; Giroux & Giroux, 2004; Kerr, 1963/2001; Kezar, 2005; Slaughter & Rhoades, 1996) during an era of conservative modernization (Apple, 2006). Public institutions are mirroring aspects of for-profit online institutions, dining halls often moonlight as catering businesses, summer camps are stuffed in residence halls, and faculty compete increasingly for external dollars tied to market-related research (Slaughter & Leslie, 1997). In conjunction with these pressures, educational equity issues have been devalued in policy discourse in order to focus on economic worth and rationalize public funding for higher education (St. John, 2007; St.John & Hu, 2006).

Moreover, recent state budget cutbacks, "along with the declining share of state funding devoted to higher education, suggest that state colleges and universities have reason to be concerned about the reliability of government support" (Lee & Cleary, 2004, p. 34) and this concern grows with each budget cycle as higher education allocations will continue to decrease throughout the next decade (Jones, 2002). As Zemsky (2005) points out,

> State governments...have consistently used market forces to solve their own short-term budgetary shortfalls by driving up the prices that publicly owned colleges and universities charge. This result occurs every time the business cycle reduces state revenues and forces state governments to choose between reducing state services and increasing state taxes. What the governor and legislature rediscover at that moment is that prisoners don't pay rent, Medicaid recipients can't pay much for health care, and public schools can't charge tuition. But, thankfully, publicly funded colleges can. (p. 279)

Such influences put incredible pressure on college and university leaders for economic survival and on state legislators to create policies that increase the number of high school graduates, improve college access, and promote graduation from college in order to increase states' "education capital" and economic development. States have decreased financial support for public colleges and universities as they have expanded demands for accountability (Tierney, 2006a). This "accountability triangle" includes state priorities, academic concerns, and market forces (Burke, 2005). Some argue that each point of the triangle holds a contradictory position, where reductive accountability from the state focuses on centralization and control whereas autonomy maintains academic freedom, but others argue the constructs are negotiable (Dee, 2006).

In addition to this financial retrenchment and political directive, disparities regarding who has access to college remain. For example, Carnevale and Fry (2001) found that in 1997, nearly 80 percent of high school graduates from high-income families went directly on to higher education, while only 50 percent of high school graduates from low-income families went on to higher education. In the same year they found that 46 percent of college-age white high school graduates were enrolled in college, whereas only 39 percent of African American and 36 percent of Latina/o high school graduates were enrolled in college. However, these statistics speak nothing of the high school graduation rates for students of the same populations, where, in 2000, 77 percent of African Americans in the 18–24 age group completed high school and only 59.6 percent of Latina/os completed high school (American Council on Education [ACE], 2002). In light of these statistics, approximately 39 percent of 77 percent of all 18–24-year-old African Americans (30 percent total) and 36 percent of 59.6 percent of all 18–24-year-old Latina/os (21 percent total) were enrolled in postsecondary education[1]—a much smaller proportion than any one statistic reveals alone.

US statistics reported by the Pathways to College Network (2004) are just as compelling. They state that by their late twenties more than one-third of whites have at least a bachelor's degree but only 18 percent of African Americans and 10 percent of Latina/os have attained degrees. These statistics may change dramatically over the next 15 years when 1–2 million additional young adults will be seeking access to higher education and a large proportion of the potential students in this group will be students of color from low-income families (Carneval & Fry, 2001), albeit which institutions of postsecondary education they would have access to is not always fully addressed and may continue to perpetuate current inequities (Brint & Karabel, 1989; Hurtado & Wathington, 2001).

Further, a perceptual gap continues to exist between students across race which has a direct impact on academic and life decisions. When comparing student perceptions of their academic performance, the importance of obtaining a high GPA declines over the college years for all ethnic groups (Sidaniusw, Levan, van Laar, & Sears, 2008). In addition, discounting academic feedback and disidentification from academics increases significantly for all students, particularly for African American students. There is also a higher level of doubt about individual academic performance in African American and Latina/o students than in white students. Moreover, access to college by people from middle- and lower-income families has been sharply reduced in recent years (McMahon, 2009).

To address such concerns about college access, the government has taken a number of national initiatives such as President Obama's American Graduation Initiative (2009), which focuses on community colleges and has set goals such as redirecting $12 billion for community colleges over the next 10 years, increasing the number of students from 5 to 10 million by 2020, instilling policies and processes that make it easier to transfer (a lesson learned from the Bologna Agreement), modernizing facilities, and establishing more online classes. More pointedly, support structures and barriers that influence access to higher education continue to shift. This shift has led contemporary theorists, practitioners, and legislators to attempt to understand higher education's current role in contemporary society and how higher education may help to increase access to college during a time of economic change as well as address the world's problems: higher education and the public good.

Friedman (2008) sums up the world's problems:

> It is getting hot, flat, and crowded. That is, global warming, the stunning rise of the middle classes all over the world and rapid population growth have converged in a way that could make our planet dangerously unstable. In particular, the convergence of hot, flat and crowded is tightening energy supplies, intensifying the extinction of plants and animals, deepening energy poverty, strengthening petro-dictatorship, and accelerating climate change. How we address these interwoven global trends will determine a lot about the quality of life on earth in the twenty-first century. (p. 5)

I take the position that higher education needs to play an instrumental role in researching and addressing myriad issues facing the world today in order to live each institutional mission and participate as conscientious community members in a diverse democracy. In this way, higher education may support the "quality of life on earth" for all, not just a select few. The importance of sincere collaboration across community-university partnerships to address problems cannot be stressed enough (Bringle & Hatcher, 2002; Fitzgerald et al., 2010; Galura et al., 2004; Pasque, 2010; Thomas, 2004; Weerts & Sandmann, 2008; White, 2005). In this sense, the pressure on higher education is twofold: (1) to tackle innumerable issues confronting students, institutions, and the system of higher education and (2) to work collaboratively with local and global communities to address complex issues including heath care, the environment (land, air, and sea), incarceration rates, drug and human trafficking, educational and economic inequities, food and water sustainability, and other issues of disparities and social justice.

In order to attempt to address the complexities of these deep and connected topics, conferences and seminars have been held across the country designed to gather leaders together to discuss the future of the relationship/s between higher education and society (for example, gatherings have been sponsored by the American Association of Colleges and Universities [AACU], 2002; AACU, 2006; American Council for Education [ACE], 2006; American Federation of Teachers, Higher Education, 2009; Association for the Study of Higher Education [ASHE], 2006, Campus Compact & AACU, 2006; Council of Graduate Schools [CGS], 2008; Department of Education, 2006; Kettering Foundation, 2008; National Forum on Higher Education for the Public Good, 2002; National Association for Equal Opportunity in Higher Education [NAFEO], 2009; State Higher Education Executive Officers [SHEEO], 2009; W.K. Kellogg Foundation, 2002). Such gatherings often present a paradox for higher education leaders interested in addressing complex issues; leaders have the responsibility to speak to a relatively small number of influential leaders about large constituencies that may or may not be represented in these small groups within a space that is not necessarily reflective of the majority of people's lives.

As an example, if you happen to be invited to the prestigious Wingspread Conference Center sponsored by the Johnson Foundation, you will be escorted by a chauffeur from the airport to the conference center through an unassuming small town. As the iron gates part to admit you onto the Wingspread grounds, the green prairie sprawls out on both sides of the winding drive. The drive is intentionally lined with pines, hardwoods, shrubs, and flowers, all perfectly manicured.

You breathe in the aroma of Lake Michigan as you round the corner to view the home that architect Frank Lloyd Wright crafted for the Johnson family (SC Johnson & Son). In the neighboring guesthouse where you will stay for the next three days, the scent of burning hickory permeates the air as the fire crackles at decibels just below recognition. The staff kindly checks you in as they reflect the caliber of hospitality for which Wingspread is known.

In your single room, you watch the sunset from large windows, sleep on pillows of feathers and sheets of Egyptian cotton, and get lost in the luxurious bathroom. The shared space of the living room is floored in warm Brazilian cherry. Books flank the fireplace of limestone, and the opposing wall is made of glass with French doors that open onto a terrace looking out to a lazily flowing river. Windows flow from floor to ceiling so you feel as if you are "in" nature, as opposed to observing nature. There is a short walk to the conference meeting house and another short walk to the former Johnson family home. You tour the various rooms of the family home, dine on a lavish meal, and even sit in the location that was a favorite of Eleanor Roosevelt when she was an overnight guest.

It is in this exquisite space where you will serve as a leader to voice issues critical to the future of higher education and the public good. Here exists the lived paradox; you are surrounded by elegance and hospitality as you wrestle with the difficult issues facing higher education and society. You must intentionally speak about how to address important community issues and at the same time not disconnect yourself from disparities around the world. From a different lens, this luxurious space provides an opportunity for leaders to focus on the critical issues at hand, rather than on their own hierarchy of needs (Maslow, 1943). This venue recognizes and elevates complex discourse while it provides a space where leaders may wrestle with important issues in order to make substantive change—an important mission, indeed.

This book focuses on the complex dialogues surrounding such paradoxes in higher education and society. Specifically, I consider the dynamic discourse between leaders who come together to discuss critical issues in higher education and the public good, albeit not at the Wingspread Conference Center venue. In the *Archeology of Knowledge*, Foucault (1976) describes that whoever holds the power regarding what counts as knowledge also has power over policy, systems, access to education, and other social processes. In the field of higher education, it is university presidents, legislators, faculty, administrators, funders, and national association researchers who hold knowledge around higher educations' multiple relationships with society and are the leaders in the field. Further, people often accept what leaders say as truth and allow them to be spokespeople for such truth (Johnstone, 2002). For these reasons, this book concentrates on higher education leaders who gather in important venues, who may or may not consider themselves gatekeepers for the field, but who do hold knowledge about higher education's multiple relationships with society.

Higher education leaders who engage in these ongoing discussions about higher education's responsibilities to society come to the conversation with competing visions, frames of reference, and worldviews (Bolman & Deal, 2008). In addition, each frame has a different set of ideas, assumptions, and implications for the continuation or interruption of current paradigms in research and policy. These leaders (legislators, university presidents, national association leaders, foundation officers, faculty, graduate students, and administrators) often talk about higher education's responsibility to serve society in extremely different ways and may—intentionally or unintentionally—labor against each other. In addition, leaders often talk about "society," the "public," or "communities" as abstractions, rather than providing specific inclusive or exclusive definitions about who or what they are talking about (Pasque, 2005). Yet, if legislators, policymakers, and the public are unclear about why higher education is important to society, then other public policy priorities may gain support at the expense of higher education (Kezar, 2004).

Throughout this book, I argue that leaders cannot afford to be complacent in this climate of educational inequity and let dominant arguments about higher education prevail. Uncovering various visions of higher education's relationships to society is paramount during this time of dramatic change. If a more thorough understanding of myriad perspectives is not offered, then dominant communicative models shared in academic discourse genres may continue to perpetuate the current ideas of higher education's relationships with society—from solely an economic rationalization perspective—without consideration of alternative perspectives that may be useful in addressing critical issues and in/equities.

By understanding more about various perspectives—or frames—and the tensions created between these frames, leaders are able to see more of the perspectives available and make more informed choices about how to work toward systemic and equitable change. In essence, it becomes imperative to view multiple frames shared through language in order to pull forward the strong points of one or more ideas or to strengthen arguments for effective policy and action. Just as we should not permit the military to serve as our only resource when working toward peace across the globe, we cannot let one perspective serve as our only option when confronting critical issues and the public good. In addition, by viewing the same policy or action through multiple frames, we may consider whether we are truly enacting equitable and just policies and actions for *all* people in society or for a select few.

Specifically, the goal of this book is to explore various leaders' competing frames of reference and worldviews of higher education's relationships with society as found in the literature (193 research articles and speeches by prominent university presidents, policymakers, and scholars) and vocalized during hours of conversation at a national conference series (four, three-day national policy conference series, behind closed doors) in order to increase our understanding of the issues and implications of various perspectives. The book is designed as a tool for current leaders interested in exploring various conceptualizations of higher education and the public good, furthering their own perspectives and working to intentionally connect knowledge,

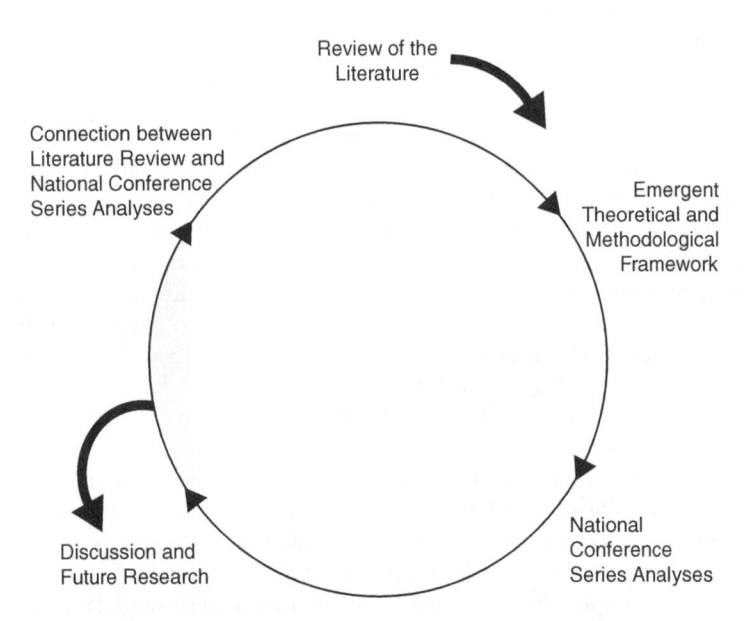

Figure 1.1   A Visual Representation of the Iterative Research Process

discourse, and action. My hope is that this information will help illuminate numerous perspectives on complex and changing issues, increase communication between leaders in various roles, influence policy decisions, and inform us so we may create equitable, idiosyncratic, and systemic change in the field of higher education.

In the remainder of this introduction, I provide an overview of each chapter. Each chapter builds upon the last in an iterative manner, furthering the depth and breadth of this analysis. See figure 1.1 for a visual representation of the iterative research process.

INTRODUCTION TO CHAPTERS

In *On Method and Hope*, Tierney (1994) mentions that research is meant to "struggle to investigate how individuals and groups might be better able to change their situations" (p. 99). I did not initiate this research study with a specific change in mind or envision that the end goal would be one of critique for critique's sake. Instead, I approach this study with the intent that emergent findings may help leaders make sense of current perspectives and this, in turn, may help us explore the implications of current prevailing and alternative frames of educational in/equity. This is similar to what Friere (1973) terms "conscientization," the knowing of reality in order to better inform it. It is with the goal of knowing, in order to make substantive equitable change, that I offer this detailed and challenging analysis about the contemporary complexities of higher education and the public good.

Specifically, I explore the cognitive processing models of higher education leaders as found in the current literature and discourse. In this study, cognitive processing models are the ways in which leaders communicate their perspectives, or frames, about a topic. This analytical process considers the conceptual framework of the participants as it emerges from their situated discourse (Taylor, 2001). This approach is based upon the assumption that people learn socially, develop a cognitive process around a construct, and then use language to show how the construct has been reconstructed. For example, in the pilot study that informed this research study, Pasque and Rex (2010) consider the various cognitive processing models offered by higher education leaders as they talk about "higher education for the public good." We state, "if we are to understand the role and purpose of higher education in a changing society, we are obliged to listen and observe what people say it is by virtue of their words and actions" (p. 2). We discern a number of cognitive processing models presented by leaders and represent the complexities of these models through an analytical framework. It was evident through the analyses that leaders' identities, memories, attitudes, and emotions played an important role in defining the nature of the problems in transforming higher education for the public good and in arriving at solutions for those problems.

The current study learns from and expands upon the pilot study. As the current leaders often refer to leaders and scholars of the past, I situate in chapter 2 the various perspectives in this book within a brief historical context of higher education and the public good. In addition, I explore 193 research articles as well as theoretical writing and speeches of higher education leaders and present a typology of the frames of higher education's relationships with society. This detailed literature review and analysis helps uncover the nuances of the various perspectives and definitions of today's leaders. For example, a number of university presidents utilize the concept of higher education and the public good as a part of their presidential platform; these include chancellor of Syracuse University Nancy Cantor (2006), president of the University of Pennsylvania Amy Gutmann (2008), and president of Wagner College Richard Guarasci (2009). I explore various leaders' conceptualizations and offer the supportive evidence for each frame, so readers may understand the tenants of each perspective. I also offer a critical analysis of the dominant and marginalized frames in higher education and discuss the relationships among frames. For instance, I describe the ways in which "human capital" is defined by leaders, some of whom use the same language and yet have competing definitions of the term. This second chapter serves as a macro-analysis of the complex linguistic dynamics of various leaders.

In chapter 3, I share the research design for this study and the importance of resisting theoretical determinism (Lather, 2003). I discuss elements of trustworthiness, including triangulation, member checking, and researcher reflexivity. In addition, I describe the specific research methods that were used in the three major aspects of this study: (1) the macro-analysis of the literature, (2) the micro-analysis of the discourse through conversation analysis and communication theories, which helps me locate the phenomena to study in more detail, and (3) the micro-analysis of the discourse through a critical discourse analysis approach, which provides added depth to this critical analysis and discussion. I describe the site for further exploration of higher education leaders' perspectives as offered behind closed doors—a national conference series (four three-day conferences) on the topic of higher education's relationships with society. In this context, the team of "higher education leaders" includes over 200 university presidents, legislators, administrators, faculty community organizers, foundation program officers, national association directors, graduate students, and a few undergraduate students.

Chapter 4 provides information on various communication theories and offers findings from a conversation analysis in order to locate the phenomena for further study. This face-to-face discussion provides a useful and natural site to explore further the discourse of the leaders on this topic from a micro-perspective of analysis. As Sacks (1964) notes, "the trouble with [interview studies] is that they're using informants, that is, they're asking questions of their subjects. That means that they're studying the categories that Members use…they are not investigating their categories by attempting to find them in the activities in which they're employed" (p. 27). This study considers naturally occurring spoken and written discourse in the current sociopolitical, economic, and cultural contexts.

Chapter 5 offers the emergent findings of what is said behind closed doors of national policy discussions regarding how to make change in our move toward higher education for the public good. The critical discourse analysis findings are quite compelling as they uncover the various dominant narratives and marginalized perspectives in these national discussions. Direct quotes from higher education leaders accompany a detailed analysis of the language used in order to reflect the natural progression of the discussion. Specifically, I explore the content (what) and process (how) of communication between higher education leaders in the hopes that both may be instructive.

In chapter 6, I dive deeper into the analysis and focus on the voices of women who resist dominant ideologies and share "advocacy" perspectives, women whose perspectives were reframed, redefined, and/or silenced in the discourse. In the analysis, I explore the content of what policy leaders missed by relegating these perspectives to the margins and how these perspectives may provide useful strategies for change. In this chapter, I focus the micro-analysis more intentionally on feminist theory, answering my own call in chapter 3 for depth of analysis through the use of one theoretical perspective after pulling from many. My hope is that this adds needed depth to the breadth of the findings from preceding chapters.

Chapter 7 offers an emergent Tricuspid Model for Advocacy and Educational Change, derived directly from the findings of this study and the voices of leaders themselves. This model focuses on organizational behavior and discourse. If such an organizational model is employed, then discourse containing concrete ideas in terms of *how* to make change in higher education may be centered and further developed. Multiple options are crucial as higher education leaders address myriad critical issues and the public good.

The concluding chapter revisits the multilevel problem that is consistent throughout the iterative research study findings and described throughout this volume. It addresses the dominant conceptions of the public good offered by leaders and urges for alternative perspectives in order to enact needed change in the immediate future. The dominant perspective pushes toward individual benefits of higher education and how these benefits may help the local and national economy. This view is important; however, it excludes notions of in/equality across social identity (i.e., race, gender, nationality, class) and does not adequately highlight the inequalities in US society and across the globe. Alternative perspectives that *interconnect* public and private good perspectives— perspectives that address economic injustice, historic and contemporary cyclical oppression, and the inherent power of the market as described in earlier chapters—are revisited. Critical issues and public good are vitally important particularly given the era of conservative modernization, privatization, accountability, and marketization of higher education today. This era, coupled with the current higher education focus on "public agenda" discourse, actively works to perpetuate dominant paradigms of stratification across class, race, gender, and other social identities (Gildersleeve, Kuntz, Pasque, & Carducci, in press). This chapter also explores what I term *parrhesiastes* of the *agora*, or speakers of dangerous truths about the interconnections between the public and private good, and about the ways in which leaders may make a concerted difference in educational research, policies, and processes, as well as in students' lived experiences.

As I explore the discourse of higher education leaders through this macro- and micro-analysis, it is important to point out that research is necessarily political (Battiste, 2006). It is political in that the current system of higher education includes rewards that benefit specific people and excludes people who have traditionally been marginalized. As researchers address what is being said and what is not being said about higher education and the public good, educational inequities, and who is included and who is excluded by the prevailing discourse, the status quo is threatened. In this sense, the various approaches described in this book are necessarily political. The discussions, decisions, and policies by higher education leaders are vital as we determine how to connect knowledge and action, the implications of which impact people's daily lives.

CHAPTER 2

Higher Education for the Public Good: A Typology

In this chapter, I explore the historical and contemporary discourse (verbal and written) about higher education's relationships with society in order to uncover current dominant cognitive processing models—or frames—on the topic. Lakoff (2006) recognizes frames as "the mental structures that allow human beings to understand reality—and sometimes to create what we take to be reality" (p. 25). First, I offer a brief overview of the conceptualizations of higher education's relationships with society from historical perspectives, as contemporary leaders often refer to earlier leaders when sharing their current perspectives. This review is not meant to be exhaustive but offers a concise temporal exploration of various perspectives of issues that continue today.

Next, I offer a typology of contemporary cognitive processing models as derived from the literature review of 193 articles. In this section, I describe the full scope of the literature review. In addition, I share each cognitive processing model, offer a few examples that represent each frame, and share the benefits and evidence that support the frames. This typology is helpful to understand the scope of perspectives on the relationships between higher education and society. There are four cognitive processing models, or frames, that emerged directly from the literature review: *Private Good, Public Good, Public and Private Goods as Balanced*, and *Public and Private Goods as Interconnected and Advocacy*.

Finally, I explore the relationships (or the lack thereof) of these various frames with society and, importantly, discuss the implications of these relationships. Specifically, I offer four analyses that further the conversation about higher education for the public good by discussing general observations and implications of the contemporary typology. The benefits of higher education, concepts of social justice and equity, conflicts of visions, contradictions in language, and definitions of "the public" are discussed.

CONTEMPORARY SCHOLARS HEARKEN BACK: A BRIEF HISTORICAL CONTEXT

Today's leaders routinely borrow from previous scholars in their own constructions of higher education's relationships with society. In some ways, these conceptualizations have not changed much over time—however, the sociopolitical context in which the perspectives are situated continues to change dramatically. In this brief section, I share some of the perspectives from scholars throughout history (prior to 1980) in order to provide a context for the conceptualizations of today.

The relationships between higher education and society have been a topic of discourse in the United States since the founding of Harvard College in 1636. As the number of colleges expanded over the years, they were viewed as a public investment and a way to educate leaders (Rudolph, 1962/1990). A discussion about higher education's role in, and responsibility to, society accompanied each catalyst for expansion. Examples include the Morrill Act of 1862 and 1890 to establish and sustain land grant institutions, the GI Bill near the end of WW II to educate returning veterans, the Truman Commission that helped initiate the community college movement in the 1950s, a commitment to a new form of civic engagement that accompanied the civil rights and social activism movements of the 1960s and 70s, the Higher Education Act of 1965 and Amendments of 1972 concerning student financial aid, and Title IX that provides legal enforcement to increase access to higher education for people of color and women (see Cremin, 1988; Geiger, 1999; Rudolph, 1962/1990). Each expansion purported to serve the public good in different ways (such as through an increase in veteran's access to higher education) albeit with different sociohistorical contexts and constraints.

In *Scholarship for the Public Good: Living in Pasteur's Quadrant*, Ramaley (2005) reminds us that the struggle for shared goals of higher education goes back to contested values addressed by the ancient Greeks; some of our current conceptualizations are not new. For example, the Sophists believed that education was to prepare a person, through rhetoric and other skills, to engage in public affairs, while Plato and Socrates believed that education ought to guide the student in understanding Truth and Beauty in the human experience.

Parker (2003) rouses Aristotle when constructing his ideas of the role higher education should play in society. Aristotle stated, "not being self-sufficient when they are isolated, individuals are so many parts all equally depending on the whole which alone can bring self-sufficiency" (as cited in Parker, p. 3). In his writings, Aristotle privileges education in order to create and sustain his notion of this "whole" of public society. Saxonhouse (1992) also shares debates from ancient Greece to highlight the influence of diversity on the capacity for democracy. She argues that Aristotle alone advanced a political theory where democracy and unity in the public world are achieved through difference.

Castoriadis (1997) also refers to Aristotle, stating that Aristotle was one of the few classical thinkers about democracy to explicitly talk about three spheres of human activities: the *oikos* (private), the *ekklēsia* (public), and the *agora* (the overlapping of the *oikos* and *ekklēsia*). The *oikos* consists of the family household, a domain where, in principle, political power should never intervene. The term *ekklēsia* is the site of political power, or "the public/public domain" (p. 7). The *agora* is the public meeting place comparable to the Italian *piazza*. This domain is where community members come together freely to discuss business matters, establish contracts, buy books, and perform other daily activities. The *agora* is the intersection of the public and private spheres. Pitkin and Shumer (1982) further our understanding of this ancient Greek tradition by stressing the duality of the public and private realms. They state,

> The public realm meant freedom, the opportunity for action, individuality, the pursuit of glory, and relations of mutuality among peers. The private, household realm was only a means to public life. It meant necessity, production to satisfy bodily needs, shame, and absence of individuation, and relations of hierarchical domination. (p. 45)

In addition to referring to the ancient Greeks, policymakers in the US education system also include conversations about "who" should have access to education. For example, some early scholars felt higher education should be available to every recognized citizen, while others felt it should be provided only to the elite. Barber (1998) reminds us that for Thomas Jefferson, literacy was the "keystone in the arch of education" (p. 170). He further adds that in 1787 Jefferson stated that the only way to preserve liberty is to "educate and inform the whole mass of the people" (cited in Barber, 1998, p. 183). Here, the strength of the American public good lies with education; however, it is important to note that in Jefferson's time references to the "masses" excluded people of color and women.

Also in the eighteenth century, Adam Smith (1776/1900) introduced the concept of the "invisible hand" whereby the public good is achieved most efficiently through individuals acting in their own, private interest. Smith's invisible hand theory is utilized today as an argument for private, individual benefits leading to the public good (Survey: Profit and the Public Good, 2005; also see Harcleroad, 1999) or as an example of why higher education should not succumb to the pressures of marketization (Barash, 2004). The contested values of higher education have continued over the years. In 1802, Joseph McKeen (as cited in Rudolph, 1962/1990), the first president of Bowdoin College, shared his perspective that

> it ought always to be remembered, that literary institutions are founded and endowed for the common good, and not for the private advantage of those who resort to them for education. It is not that they may be able to pass through life in an easy or reputable manner, but that their mental powers may be cultivated and improved for the benefit of society. If it be true no man should live for himself alone, we may safely assert that every man who has been aided by a public institution to acquire an education and to qualify himself for usefulness, is under peculiar obligations to exert his talents for the public good. (p. 58)

McKeen's perspective was the antithesis of the view expressed by the influential University of Michigan president Henry Tappan, who complained in 1850 that "we have cheapened education so as to place it within the reach of everyone" (as cited in Rudolph 1962/1990, p. 63). Debates about who should have access to education in order to serve the public good remains contested today (Bloom, 1987; Bowen & Bok, 1998; Brint & Karabel, 1989; D'Souza, 1991; Giroux & Giroux, 2004; Green & Trent, 2005; Hagedorn & Tierney, 2002; Levine, 1996; Lewis, 2004; Schlesinger, 1998).

The connection between the public and the private good was also prevalent in the founding of some external enterprises. For example, Andrew Carnegie, a steel baron and philanthropist, and Henry Smith Prichett, a scientist and educator, collaborated on the founding of the Carnegie Foundation in 1905. Carnegie and Prichett's perspectives on philanthropy have been serving as the basis for the organization and, in part, for the field of philanthropy today. They believed that "private power could be [used] for the public good" (Lagemann, 1983, p. 2), a belief that encouraged a strong association between higher education and the wealthy.

In a 1928 speech, the then president of the University of Minnesota, Lotus D. Coffman, emphatically stated "The progressive advancement of democratic institutions depended upon an educated citizenry" (cited in Brothen & Wambach, 2004, p. 8) and encouraged increases in access to the university in order to reap benefits for industries across the state. The connection between higher education and training workers for industrial fields such as engineering, science, and technology has been increasing since the 1950s (Kerr, 1963/2001).

The *Brown v. Board of Education of Topeka* decision in 1954 also left a fundamental mark on the relationship between higher education and society. The Warren Court was able to unanimously say what the Vinson Court could not—that separate but equal was unconstitutional. This decision was primarily focused on elementary and secondary schools, but it also had an impact on postsecondary education. The decision was contested and equitable access to higher education continues to be legally challenged even today.

More recently, in the 1970s, institutions of higher education were told to measure the effectiveness of their outputs and to justify their costs (Lawrence, Weathersby, & Patterson, 1970), another trend that continues even today. A conference on the topic of outputs of higher education was organized to further explore public and private benefits of higher education, as well as to measure these benefits. Issues such as public accountability, the inputs and outputs of higher education for public and private good, and how to measure these outputs were discussed. However, each of the speakers defined benefits and outputs differently, some in terms of the public good, others in terms of the private good. Different conceptualizations and approaches for research on the role of higher education in society were explored and none emerged as dominant.

One of the most comprehensive conceptualizations of higher education's relationships with society is in Bowen's (1977) book *Investment in Learning: The Individual and Social Value of American Higher Education*. Bowen considers individual benefits produced by the academy, such as "personal development, economic opportunity, rich satisfactions," and societal benefits including "political, economic, and cultural advancement" (p. xiii). Bowen found that higher education affects individuals through cognitive learning, emotional and moral development, citizenship, economic productivity, consumer behavior, leisure activities, and health. The impact on society included educational outcomes such as changes in college education, individual participation in political leadership and transmission of values to their own communities, and progress toward human equality. After contemplation of the public and private benefits of higher education, Bowen concluded that a balanced perspective was needed. He stated,

> The monetary returns alone, in the form of enhanced earnings of workers and improved technology, are probably sufficient to offset all costs. But over and above the monetary returns are the personal development and life enrichment of millions of people, the preservation of the cultural heritage, the advancement of knowledge and the arts, a major contribution to national prestige and power, and the direct satisfactions derived from college attendance and from living in a society where knowledge and the arts flourish. (p. 447)

The connection between higher education and society has been discussed throughout history, albeit how it was defined and what social milieu influences each author differ. In addition, countless other catalysts (beyond the scope of this brief historical context) of expansion and publications have influenced contemporary leaders (Smith, 1776/1990; Becker, 1964/1993; Buchanan & Tullock, 1962; Dewey, 1916; Freire, 1970/2002; Keynes, 1936). The implications of each perspective, law, and educational policy have direct implications for people throughout US society. With the addition or recycling of each perspective, the systemic and institutional perpetuation of inequity in education continues. We have yet to interrupt the cycle of oppression and educational inequity in American higher education.

A Contemporary Typology of Higher Education for the Public Good

In this section, I describe the scope of the literature review and analysis. In addition, I share each of the four cognitive processing models *(Private Good, Public Good, Public and Private Goods as Balanced,* and *Public and Private Goods as Interconnected and Advocacy)*, offer a few examples that represent each frame, and share the benefits and evidence that support the frames.

In order to explore contemporary literature, the following orienting questions were asked: What are the different contemporary conceptualizations of higher education's relationships with society that have been developed by higher education leaders? What are the benefits (and evidence for these benefits) of higher education for society (or for individuals who may then benefit society) that are associated with the various conceptualizations? And, importantly, what are the implications of the relationships between conceptualizations?

For this study, I reviewed 193 contemporary (1980–2009) articles, books, reports, and speeches that mention higher education's relationships with society in higher education, business, economics, K-12 education, policy, political science, psychology, and sociology. A theoretical sampling process (Strauss & Corbin, 1999) was used where sources were gathered from an extensive and systematic search of library databases, relevant websites, course syllabi, conversations with colleagues immersed in the topic, and select references from relevant articles (Hart, 1998). Themes were inductively compared across articles (Strauss & Corbin, 1999). I created a table as a tool for analysis that changed over the course of the collection of articles to reflect the themes that emerged from various sources; the typology of higher education leaders' frames presented here emerged from this table.

In order to delimit the research, I concentrate on articles that talked about higher education as a (or "for the") public or common good, or about public benefits of higher education. This analysis specifically excludes perspectives that view higher education as only a private benefit—a public benefit that is solely a by-product of a private, individual benefit and *not* a public good (Friedman & Friedman, 1980; See Bloom, Hartley & Rosovsky, 2006).

Each of the four cognitive processing models is described below along with a few examples and its benefits and evidence.

The Private Good

The Relationships between Higher Education and Society

The literature that explores higher education's relationships with society as a public good through investing in the private individual is extensive. The primary authors who hold a *Private Good* frame include economists, policy scholars, legislators, government agencies, and leaders—all vying for limited dollars (e.g., Bartik, 2004; Becker, 1964/1993; Brandl & Weber, 1995; Gottlieb & Fogarty, 2003; Small Business Association, 2004; Weiss, 2004; Weissbourd & Berry, 2004a, 2004b). These higher education and business leaders' vision is that educating the private individual will contribute to the public good through an increase in economic growth; they thereby define the public good as local, state, and national economic vitality. Positive externalities, monetary or nonmonetary spillover benefits to others, are experienced but not often realized by those who invest in education (McMahon, 2009).

The primary argument is to sustain resources such as continued government subsidization of colleges and universities so that individuals may participate in higher education, an effect that will in turn influence the public good. This is reminiscent of the traditional input-output model, which emphasizes that educating individual people (input) would lead to these individuals working to increase the national, state, and local economies (output). As Paulsen and Toutkoushian (2007) describe, "economists think of colleges in much the same way as other organizations in that they rely on an input-production-output to deliver higher education services" (p. 19).

In this sense, this frame is different from the Friedman and Friedman (1980) perspective as the benefits to society are public and private, albeit delivered from private goods. In addition, economic rationalists believe that the national economy will suffer if higher education does not privatize research to protect its own interests (Brown & Schubert, 2000; Currie & Newson, 1998). In this conceptualization, higher education is the "engine of growth" for the economy (Becker & Lewis, 1993) and there exists a substantial dialogue among and between constituencies who hold this perspective, often directed at policymakers as the primary audience.

One example of this perspective is from the National Center for Public Policy and Higher Education (NCPPHE) (2003), which argues that individual progress achieved through higher education benefits all states. They urge legislators and policymakers to make financial and policy change in concert with these beliefs for the good of the state and national economy. NCPPHE states,

> Individuals with higher degrees can expect to earn higher incomes. The result: more tax revenue and economic activity for the state. An educated, skilled population makes fewer demands on social services such as welfare and corrections. The result: less expense to the state. People with more education make more informed health and life style choices. The result: state savings in public resources. Educated individuals are more comfortable handling decisions about health care, personal finance and retirement. The result: less government responsibility in those areas. (p. 1)

The NCPPHE encourages each state to invest in individuals by increasing college and university graduation rates in order to expand the local, state, and national economy as well as to reap the benefits mentioned above.

Gary Russi, president of Oakland University and chair of the President's Council of State Universities of Michigan, also has a state-focused vision for change. Russi (2004) emphasizes on investment in the individual so that individuals may make a difference to state and global economies. Russi fears that a decline in state support will put colleges and universities in jeopardy of evolving from "state-supported, to state-related, to state-located" (p. 5) institutions and encourages continued state support of colleges and universities in his state. Russi communicates that "public higher education is a public good and that the public investment made by the citizens of the State of Michigan is returned many fold and in countless ways" (p. 5). The "ways" Russi mentions are solely through an investment in private individuals, who will then contribute to the public good by economic means. For example, Russi states, "Our universities are serving as catalysts for new business start-ups, job creation, and the attraction and retention of highly educated individuals" (p. 2) and for "every one dollar of state investment in our universities generates $26 of economic impact. In other words, the state's $1.5 billion in investment in our public universities produces $39 billion in economic impact, or a remarkable 12.6 percent of Michigan's entire gross state product" (p. 3). This vision echoes the *Final Report of the Lt. Governor's Commission on Higher Education and Economic Growth* presented to Michigan governor Jennifer Granholm in December 2004. Empirical studies in the report reveal that postsecondary education fosters discovery of new ideas, prepares individuals with skills, builds dynamic and attractive communities, and creates greater prosperity for the state economy. Again, in this *Private Good* conceptualization, an investment in individuals through higher education is attached to the State of Michigan's economic vitality.

In another example, former Minnesota state legislator John Brandl and former congressman Vinn Weber, a self-proclaimed liberal and a conservative respectively, address the economic benefits of higher education in their *Agenda for Reform* (1995) report for the governor. They pose a primary question, "How can a free people regularly and dependably accomplish public purposes?" (p. 10). The answer, for Brandl and Weber, lies in competition and community. They suggest that state funding for higher education be shifted primarily to individuals themselves, so that students choose the college or university to attend and invest in, thereby creating a competitive market driven by institutional competition and student choice. In this conceptualization, institutions of higher education become less dependent on governmental support and have more of a focus on market-driven revenues gained through individual students. Brandl and Weber state, "In both private and public realms, competition is fundamental. When citizen-consumers have the choice between competing suppliers, then those individuals possess the power that holds the suppliers accountable…Competition puts power in the hands of individual citizens, not bureaucracies" (p. 10).

The major recommendation from Brandl and Weber is to "radically change the way state funds for higher education are appropriated by giving more to students and less to institutions" (p. 25). This vision of competition reallocates state appropriations for colleges and universities to individuals as it supports a market-based and competitive higher education enterprise. Specifically, Brandl and Weber recommend giving directly to individuals who seek education and training 60 percent of state appropriations in the form of learning grants and need-based grants where this "should be thought of as replacing the bulk of the current state appropriations" (p. 25). The authors reserve 30 percent of state appropriations for direct institutional support in the form of block grants to the two public higher education systems in Minnesota. The grants would not be limited to public institutions and could be used for all higher and postsecondary education, both for- and not-for-profit. Finally, 10 percent of the budget would be available for basic and applied research and for statewide programs such as the interlibrary loan system.

Brandl and Weber purport that this set of recommendations will ensure that the weakest institutions will not survive the "post-secondary education marketplace" but will enable state institutions to compete fairly with other providers by "removing the handicaps of current state administrative and regulatory structures and policies" (p. 25) so state funds do not maintain inefficient campuses. These market-driven recommendations would dramatically alter the role of the Minnesota State Colleges and Universities Board of Trustees who in 1995 controlled 65 percent of the funding that colleges and universities needed to operate. Brandl and Weber believe that the board would continue to focus on quality assurance issues; however, without a significant amount of funding as a leverage, they suggest, the board's authority comes primarily from setting standards of academic management and performance and statewide postsecondary educational goals and objectives. Audits will enable the chancellor to hold presidents responsible; however, the authors fail to provide information about the leverage the chancellor would need to hold the presidents accountable.

Brandl (1998), continuing his work on competition and community, states that it is the government's role to "arrange affairs in such a way that both private citizens and public employees can accomplish public purposes as well as acting ordinarily and freely in ways that accomplish their own objectives" (p. 1–2). In this vision, Brandl privileges individuals' actions for their own private, individual good and yet believes that it is the government's role to ensure that private good actions connect with the public good. Brandl continues with his notions of competition through individual choice and an increase in the role of community agencies. In this manner, individuals accomplish their personal objectives as the government utilizes competition and community as tools to ensure that private interests fulfill both private and public needs.

Toutkoushian and Shafiq (2007) further a similar argument, one that argues for the idea of states giving aid directly to students rather than institutions. They do not, however, factor in the benefits from competition as is found in Brandl's arguments but instead focus on how more students might be able to attend college through a more efficient and effective financial support program provided through a shift in how the state disperses its resources. In this way, their revision of the process of allocating higher education funds to individuals helps to support the public good.

Supportive Evidence

There are a number of purported benefits from the *Private Good* conceptualizations of higher education's relationship with society; these include increased US Gross Domestic Product (GDP); national productivity; income and wage rate; national, state, and local revenue; economic welfare of a city; knowledge spillovers; campus resources and space; and community resources. Scholars, particularly economists and leaders fighting for additional state dollars, mention the increase in individual income as a benefit from higher education for society.

Many of the benefits, and the evidence that supports these benefits, overlap with each other. The research is conducted at either a national level or a local city level; however, each researcher did mention that these categories are not mutually exclusive, that is, increase in wages influences federal, state, and city taxes. It is also important to note that economists themselves disagree about the importance of market forces as a factor contributing to the value of higher education (Fiorito & Kollintzas, 2004; Laney, 1981; Melissas, 2005; Soros, 2001; Sternberg, 1993).

Gottlieb and Fogarty (2003) give an example of the benefits and evidence to support these benefits from the *Private Good* frame. They state that a number of economists are concerned with the theoretical aspects of human capital and provide empirical evidence that local or national benefits of higher education exceed individual benefits. Human capital is defined as individual earnings, state and individual rates of return on investment, and national and local economic growth (Becker, 1964/1993; Blinder & Weiss, 1976; Gottlieb & Fogarty, 2003; Weiss, 1995). Specifically, Gottlieb and Fogarty found that educational level is one of the strongest predictors of economic welfare for a city. This idea justifies higher education as a value to market economies and as a societal good. Using 1980 and 2000 US Census data, they found that among 267 metropolitan areas in the United States, "an educated workforce is a significant determinant of subsequent per capita income growth" (p. 331). They also found that educational attainment (defined as the percentage of the population with at least 4 years of college) was a significant predictor of per capita income growth for local cities demonstrating 4 percent increase (p < .001) over 20 years. The authors controlled for any increase in labor force participation, specialization in manufacturing, size of the city, and region of the country. Gottlieb and Fogerty reason that this information should encourage national legislators to financially support individuals to attend colleges and universities in order to support the national, public good.

DesJardins' (2003) study focuses on the state of Minnesota in order to ascertain whether the state is making a sound monetary investment in public higher education. His study included individual income and tax revenues as factors defining the benefits of higher education. He assesses the private returns of completing a bachelor's degree (earnings) and weighs it against the private costs incurred (tuition, books, transportation, etc.). He also considers estimated lost earnings while a student is enrolled in college. He found that "the State will accrue an additional $57,018 in non-discounted income tax over the working lifetime of each [male] bachelor's degree recipient" (p. 186). In addition, alumni will spend more money in contribution to the local economy. DesJardins also found that the State of Minnesota's internal rate of return (IRR) for males who had some college is conservatively 3.6 percent, for a bachelor's degree the IRR is 8.4 percent, and for a professional degree 11.2 percent. The individual IRR is 4.9 percent, 12.5 percent, and 18.5 percent respectively. Corresponding outcomes for women proves positively significant and yet not as strong as those for men. DesJardins concludes that public subsidization of public higher education is a "win-win proposition" (p. 196) for the state.

Moreover, Day and Newburger (2002) of the US Census Bureau found a significant increase in earnings with education level. The average annual wage for high school dropouts is $18,900, for individuals with a high school degree it is $25,900, and for individuals with a college degree the figure is $45,400. Further, the National

Center for Education Statistics (Decker, 1997) found that postsecondary education training may increase an individual's weekly earnings by up to $140.

THE PUBLIC GOOD

The Relationships between Higher Education and Society

The higher education leaders who talk about the relationship between higher education and society from the *Public Good* perspective are primarily university presidents and key spokespeople for national higher education associations who state their vision for the future of educational institutions or the system of higher education (e.g., Cantor, 2003; 2007; Campus Compact, 2004; González & Padilla, 2008; Guarasci & Cornwell, 1997; Rosenstone, 2003). These scholars believe that higher education's primary role is to educate students to participate in a diverse society and contribute to society in a positive manner. Further, principles of democratic education, community, and exemplar teaching pedagogy simultaneously help educators develop students for effective civic participation in a pluralistic society.

For example, Guarasci and Cornwell (1997) state that higher education has a responsibility to society to encourage citizenship through civic education, to prepare students for a diverse democracy and to participate in the public good. They argue that institutions of higher education must do more than reform the curriculum to further democratic aims and call for system-wide revisions to hierarchical organizational strategies and compartmentalized ways of knowing and being. It is not enough to educate *for* the public good; higher education institutions must also operate *as* a public good. One way Guarasci suggests that Wagner College (where he serves as president) enact this vision is through creating service-learning programs and living-learning programs that not only provide education for students to participate in society but also connect academic and student affairs throughout the institution (Guarasci, 2009). Collaboration between faculty and staff will help the institution model a diverse democracy for the benefit of society.

A second example of this perspective is from Cantor (2003; 2007) who talks about the significant relationship between higher education and society and addresses higher education's responsibility to create diverse learning environments. In 2003, as the chancellor of the University of Illinois, Cantor challenged university community members to consider "higher education for the public good" throughout the institution. She provided financial and institutional support to programs and departments that have goals and objectives consistent with this vision of the public good. Cantor announced funding for a new center for democracy and an expansion of the intergroup dialogue program through the creation of an intergroup dialogue living-learning program for undergraduates. Cantor connected these initiatives with the university's responsibility to the local community and the state. Cantor focused on the education of students through civic engagement and multicultural education for the public good. The individual benefits of the *Private Good* frame are not included in her message.

> For this country to move together peacefully, it will not suffice to integrate the boot camps and not the military academies, the juror boxes and not the judiciary, the emergency room and not the operating theater, the factory and not the boardroom, the classroom and not the professorate, the voting booth and not the Congress. Real integration can not [sic] happen until Americans of all colors learn with and from each other in the best classrooms of this land and thereby position themselves for leadership. (p. 4)

Later as the chancellor of Syracuse University, Cantor furthered her ideas of higher education as a public good and made this the crux of her platform for the future of the university. In a 2007 speech, Cantor stated, "reassessing the public benefits of higher education is one of the biggest challenges of both public and private universities" and institutions may do this through public scholarship. This private scholarship (local and global community-university partnership) is what "offsets the private gains and moves diversity and diverse culture to the center of the campus community."

Another example of the *Public Good* perspective is from Rosenstone (2003), a political science professor and dean of the college of liberal arts at a large public research university who specifically argues against the private good. He contends that higher education administrators do not need to communicate the economic benefits of universities but should educate legislators, business, and civic leaders to recognize the importance of research and how the dissemination of knowledge serves the greater public good. He further asserts that universities need to rededicate themselves to their core principles, realize that no university can do it all (i.e., they should determine disciplinary specialization/s), educate policymakers and the public about the idea of the university as a public good, and enhance research creativity.

González and Padilla (2008) reflect that "in Hispanic culture, the tradition is for institutions of higher education and the people who pass through them to enhance the public good through their technical competence and their presumed superior moral development and character" (p. 5). Such a tradition may be instructive as universities work to restore their civic mission in a way that supports the public good.

The concept of social capital as defined by Putnam is mentioned in many of the *Public Good* conceptualizations. Putnam (1995) states,

> by analogy with notions of physical capital and human capital—tools and training that enhance individual productivity—social capital refers to features of social organization such as networks, norms, and social trust that facilitate coordination and cooperation for mutual benefit. (p. 67)

The decline of social capital is a theme of Putnam's work (1995; 2001), yet there may be interventions to increase social capital. One of the central factors in the decline of social capital is television, which is seen as having a profound privatizing impact that undercuts social capital in a society (Putnam, 1995). Further, it is Putnam who devised a social capital index utilizing national data sets including such factors as community organizational life, engagement in public affairs, volunteerism, informal sociability, and level of social trust in a state's population. An increase in these social capital factors will enhance the public good, this theory is similar to Cantor and Guarasci's argument for an increase in civic engagement and participation in a diverse democracy. Principles of Putnam's social capital are found in the education initiatives and ideas that Rosenstone and other leaders describe.

Supportive Evidence

A number of qualitative, quantitative, and mixed-method studies support educating students to participate in society for the public good (Kerrigan, 2009; Markus, Howard, & King, 1993; Maxwell, Traxler-Ballew, & Dimopoulos, 2004; Perry & Katula, 2001; Rowley & Hurtado, 2003). One highly cited study considers student participation through service-learning (Astin et al., 2000). Astin et al. conducted a mixed-method longitudinal study that included 22,236 undergraduates throughout the United States. The quantitative impact of service-learning was assessed on eleven different dependent measures and controlled for institutional and student characteristics. Participation in service-learning showed significant positive effects on all eleven outcome measures, including academic performance, values, self-efficacy, leadership, choice of a service career, and plans to participate in service after college. The qualitative aspect of the study involved in-depth case studies of service-learning on three different campuses. The qualitative findings suggest that service-learning is effective as it facilitates an increased sense of personal efficacy, awareness of the world, awareness of one's personal values, and engagement in the classroom. Astin et al. also found that both faculty and students "develop a heightened sense of civic responsibility and personal effectiveness through participation in service-learning courses" (p. 5). Together, the quantitative and qualitative findings suggest that "providing students with an opportunity to process the service experience with each other is a powerful component of both community service and service-learning" (p. 3). The study also found that undergraduate participation in service-learning increases civic engagement after college.

A second example of evidence for a public good perspective of higher education is from Gurin, Dey, Hurtado, and Gurin (2002). This example considers student participation in a diverse democracy during and after the time students are in college and was used, in part, as evidence for the defense in the *Grutter v. Bollinger* (2003) and *Gratz v. Bollinger* (2003) US Supreme Court cases on affirmative action. The authors identify patterns of educational benefits on the individual institution level and across institutions. The researchers utilized two longitudinal databases. The institutional database included 1,129 white students, 187 African American Students, and 266 Asian American students. The national database (Cooperative Institutional Research Program) included 10,465 white students, 216 African American students, 496 Asian American students, and 206 Latino/a students. Students were surveyed the first year and again four years later. The study controlled for ethnic/racial composition of the high school and precollege neighborhood, gender, high school GPA, SAT, parents' educational attainment as a measure of the student's socioeconomic background (SES), and institutional features.

Gurin et al. (2002) use structural diversity (the number of people from diverse groups), informal interactional diversity (the frequency and quality of intergroup interaction), and classroom diversity (learning about and gaining experience with diverse people) as factors in this study. Specifically, the authors examined the relationship between these different types of diversity and four dependent variables (intellectual engagement, academic skills, citizenship engagement, and racial/cultural engagement). Multiple regression analyses determined that in the national study, informal interactional diversity and classroom diversity explained between 1.5 percent and 12.6 percent of the variance in the different educational outcomes for the four [racial] groups. In the Gurin study, the three diversity experiences explained between 1.9 percent and 13.8 percent of the variance across the educational outcomes of the three [racial] groups. (p. 358)

Outcomes were significant (p < .05 or less) for each racial group. The findings show that the "actual experiences students have with diversity consistently and meaningfully affect important learning and democracy outcomes of a college education" (p. 358). Based on the results, the authors argue for more structural, classroom, and interactional diversity on college campuses.

PUBLIC AND PRIVATE GOODS: A BALANCED FRAME

The Relationships between Higher Education and Society

The scholars in this *Public and Private Goods: A Balance* (or *A Balanced*) section acknowledge both the public and the private good benefits to society as previously defined. The authors are typically policy analysts or researchers at national higher education associations (e.g., Baum & Payea, 2004; Boulus, 2003; Callan & Finney, 2002; Institute for Higher Education Policy [IHEP], 1998; IHEP, 2005; McMahon, 2009; Wagner, 2004). Scholars with this view urge for a "both/and" model where higher education is both a public and a private good. The authors from this frame do not, however, address interconnections between the public good and the private good; the two aspects may influence one another, yet each entity is described as separate from the other. This public and private goods argument is most often utilized to extend the benefits of higher education beyond the private good and argues for continued support or assessment of higher education. This frame is a compromise between or a balancing of the influential private good perspective and the communal aspects of the public good perspective.

The Institute for Higher Education Policy's (1998) Array of Benefits best represents this conceptualization and it is often cited by leaders in the field of higher education (See figure 2.1). The Array of Benefits is a figure that adds the oppositional categories of "economic" and "social" benefit to those of the "private" and "public" good. This illustration validates the frequently discussed perspective of the "private/ economic" benefits of higher education such as higher salaries, perks, and savings. It also furthers our understanding of the role of higher education in society by including "public/economic," "private/social," and

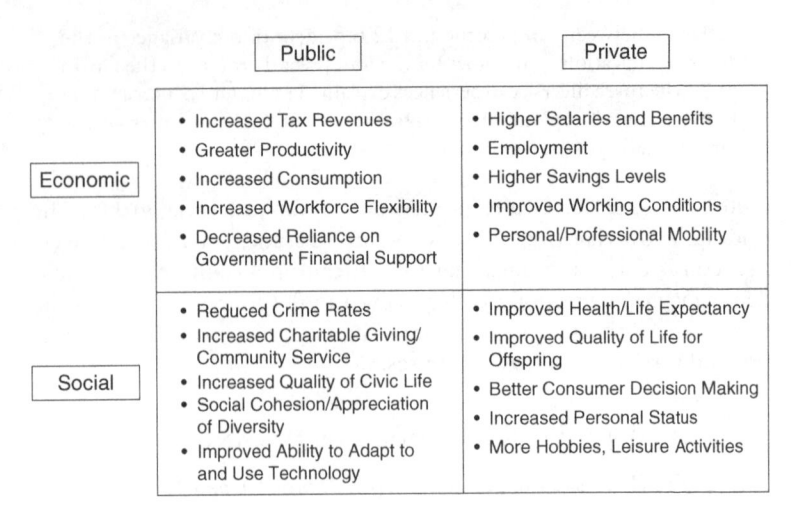

Figure 2.1   Array of Benefits

The Institute for Higher Education Policy (1998) in *Reaping the Benefits Defining the Public and Private Value of Going to College.*

"public/social" benefits. IHEP describes these four categories as mutually exclusive, although it briefly mentions that a private, economic benefit could spill over onto the public, economic benefit category. Supportive evidence for this perspective is cited by the IHEP within each category.

The IHEP recommends a more formal system for measuring and reporting the benefits of higher education in order to increase public support for higher education across the nation. This system could ensure that legislators, parents, and all stakeholders are informed about the benefits of attending college. It could also help policymakers and higher education administrators change the way society invests in higher education, from a private/economic to a public/social good, as the private/economic good is garnering too much attention. They argue that such a national reporting system will increase our understanding of the benefits of higher education and help further dialogue about the public good.

McMahon (2009) also describes this balance between the public and the private by furthering the concept of efficiency as related to externalities in serving the public good. He states that "both internal efficiency (related to unit costs) and external efficiency (how well the outcomes relate to social benefits expected by society)" (p. 12) make a difference in higher education and the public good. Further, if higher education does not address both the private and the social benefits of higher education, then the United States will continue to lose its "comparative advantage" as compared to other countries around the world, particularly South Korea, China, and Canada (p. 24).

Finally, some authors with this frame, such as Callan and Finney (2002), address the idea of "capital," but in terms of "educational capital." Educational capital (also cited as academic capital) in these contexts is defined as goods yielded by higher education that are both private (including human capital) and public (including social capital). The combination of human capital and social capital addressed in this frame can be found in the increasingly cited American sociologist James Coleman's (1988) use of the term "social capital." Coleman's theory of social capital is grounded in a structural-functionalist theoretical frame. Coleman attempts to merge sociology and economics using "the economists' principle of rational action for use in the analysis of social systems proper, including but not limited to economic systems, and to do so without discarding social organization in the process" (Coleman, 1988). Here, social capital is defined by its function to include the social structures and actions of people within the structure. This theory is utilized in numerous research studies and helps scholars identify a number of educational capital factors related to educational outcomes (for a literature review of 443 journal articles on social capital from 1986 to 2001, see Dika & Singh, 2002).

PUBLIC AND PRIVATE GOODS: AN INTERCONNECTED AND ADVOCACY FRAME

The Relationships between Higher Education and Society

The scholars who perceive the public and private relationships between higher education and society as *Interconnected* and speak with a voice of *Advocacy* all have two similarities. First, the authors discuss a mutual interdependence between public and private good; the location where one ends and the other begins is blurred. Second, the authors each passionately describe a crisis in higher education where action from leaders is needed to change the focus of the education from a capitalistic, privatized, market-driven model to one that better serves an inclusive and diverse public good in order to promote educational equity and justice. Each author explores issues of power and an inclusive definition of who should have access to college across race, gender, and/or class. The scholars who conceptualize the relationships between higher education and society—where public and private good interconnect—are primarily tenured faculty from the social sciences (e.g., Giroux & Giroux, 2004; Green & Trent, 2005; Hagedorn & Tierney, 2002; Kezar, 2005; Labaree, 1997; 2007; Nussbaum, 1999; Padilla, 2008; Parker, 2003; Pitkin & Shumer, 1982; Rhoades & Slaughter, 2004; St. John, 2003; 2006; Torres, 1998; Urrieta, 2008).

In the *Advocacy* frame, one of higher education's roles in a democracy is to acknowledge the public and private realms as well as privilege the interconnections between them. The authors view this interconnection as the crux of a crisis in the academy where change is needed in leaders' perspective and behavior vis-à-vis the academy. They believe it is particularly important for leaders who hold power within colleges and universities to initiate change in order to address educational inequity across race, gender, and class within and outside of the system of higher education. Political capital and changing to actualize a true and inclusive democracy—inclusive of all dominant and marginalized voices—are central. In a *balanced* perspective, one may consider factors that contribute to the relationship between higher education and society as mutually exclusive, whereas in this perspective, the intersections of various aspects within and outside of the academy render isolation of a factor virtually impossible.

Most scholars with the *Interconnected and Advocacy* perspective identify people with an economic neoliberal view—who support the marketization of higher education—as a serious problem and believe that the academy lacks leadership and governance. People with an economic neoliberal view are seen as people who hold power in this context. The authors fear that if there is not a change in how stakeholders with power perceive and act upon higher education's relationships with society, then higher education will be increasingly perceived as a private good, or a commodity. This will continue to perpetuate inequities in this era of conservative modernization (Apple, 2006). Conceptually, conservative modernization signifies a hegemonic bloc of social forces—or movement—that colludes to effect conservative changes in education. Solutions are often identified such as increased access to education around race, gender, nationality, and class; multicultural education and civic engagement for a diverse democracy; refinancing postsecondary education; and a change in educational leadership.

Supportive Evidence

For clarity, I have divided this complex frame into four categories based on the primary focus of the authors. For example, Parker (2003) and

Pitkin and Shumer (1982) connect the notion of civic engagement and multicultural education with a diverse democracy as a way to focus on the intersections between public and private good. Labaree (1997; 2007), Rhoades and Slaughter (2004), and Kezar (2005) focus on higher education as a marketplace and the political nature of the intersections of public and private good. St. John (2003; 2006) addresses the role of finances in promoting access and equity. The Girouxs (2004) connect all of these issues together as they address this blurred vision of the relationship between higher education and society to reduce educational inequity around race, gender, and class. However, these subcategories definitely overlap, which is indicative of this frame's resistance to finite boundaries. For example, the subcategories overlap in their emphasis on the perspectives of people in power, that educational transformation needs to be inclusive of perspectives from diverse groups, and that educational change must address race, gender, and class inequities.

Civic Engagement and Multicultural Education as the Intersection of the Public and Private

Parker (2003) addresses the relationships between higher education and society through his conceptualization of idiocy versus citizenship where the idiot is the "private, separate, self-centered" (p. 2) person and the citizen is the public actor. Parker breaks down this dichotomy by describing democratic citizenship education in terms of "an understanding of both *pluribus* (the many) and *unum* (the one), and an understanding that the two are, in fact, interdependent" (p. 1). Parker further adds that institutions of higher education have not fully grasped the connections between these two, as is mirrored in the perceived dichotomy between public and private good, and in the separation between multicultural and citizenship education for participation in a diverse democracy. Parker continues that higher education can promote the interconnection between the *pluribus* and the *unum* through multicultural education linking to democratic citizens and leading to a culturally, racially, and politically diverse society (also see Torres, 1998). These various aspects progress from one another and are also inextricably linked. The higher education-society relationship includes responsibility to the public, the private, and their intersections, where unity, as opposed to dichotomy, arises from diversity. In this manner, Parker believes that higher education needs to fulfill its responsibility to educate students for the private and the public good of society.

Pitkin and Shumer (1982) connect the public and private good, but from a perspective of radical criticism where they specifically address issues of privilege and oppression. The authors state that the "crucial function" of civic engagement in and out of higher education is to connect the private good with that of the public good. Their call is in "revolutionizing" the power of our democracy by "transforming people from consumers, victims, and exploiters into responsible citizens, extending their horizons and deepening their understanding, engaging their capacities, their suppressed anger and need in the cause of justice" (p. 48). They further state that "the idea of democracy is the cutting edge of radical criticism, the best inspiration for change toward a more humane world, the revolutionary idea of our time. The basic idea is simple: people can and should govern themselves" (p. 43). The authors are clear that the crucial function of political engagement is to connect the personal, individual good with the public good.

In his revealing contribution to an autoethnographic research project of Latina/o faculty perspectives on higher education and the public good, Padilla (2008) talks about how the public and private "can easily morph into the other" (p. 13). He describes this interconnection through his own experiences as a student, migrant worker, valedictorian, admissions counselor and organizer and, through his story, reveals the ways in which the two become easily entangled throughout one's lifetime.

The Marketplace and the Political Nature of the Intersection of Public and Private

A number of authors discuss the rise in the commercialization and privatization of research in recent years and the potential for additional collaboration between higher education and industry. In this conceptualization, however, the authors intentionally argue that this is not in the interest of the public good (Bok, 2003; Kerr, 1963/2001; Kezar, 2005; Slaughter & Rhoades, 1996).

Professor of K-12 education Labaree (1997) discusses the narrow pursuit of private advantage at the expense of the public good. He states that "by constructing a system of education so heavily around the goal of promoting individual social mobility, we have placed public education in service to private interests" (p. 261). Labaree's perspective of the current state of higher education is often mirrored in state budget allocations, as more of the cost for college is moving away from the state to the individual family. He argues that the push and pull between public and private good results in a no-win situation. Instead, Labaree argues for attention to be paid to all that is "fundamentally political" (p. 16) including the intersections of both the public and private good. Here, the interconnection of public and private is the political capital arena, which he argues is central to the education discussion.

Higher education scholars Rhoades and Slaughter (2004) describe the shift in the United States from an industrial to a knowledge- and informational-based economy. The authors state that this shift, coupled with a decrease in state funding, has served as a catalyst for an increase in the generation of income from institutional teaching, research, and service areas. There may be a transformation in the economy and a "blurring of the boundaries" (p. 38) between profit and nonprofit. This transformation comes from an increase in "neoliberal and neo-conservative politics and policies that shift government investment in higher education to emphasize education's economic role and cost efficiency" (p. 38), which, in turn, has led administrators to increase market capitalism on campus in order to survive. The authors believe that faculty have been too complacent in the face of these changes. They argue that in this political environment, college and university faculty and administrators need to be more forceful and provide alternatives to academic capitalism in this changing economy. Higher education needs to provide access to postsecondary institutions, prepare citizens for a diverse democracy, and address social problems and issues.

Rhoades and Slaughter (2004) include specific recommendations for change including increasing access to higher education and preparing citizens to engage in a democracy. They also argue for faculty, national associations, and faculty unions to "reprioritize the democratic and educational functions of the academy" to include "more public discussion and more public accountability" (p. 57) and urge them to examine who benefits from the current system and to advocate for the inclusion of people who have been historically excluded. The authors argue for a "'republicizing' of US colleges and universities" (p. 57).

Higher education scholar Kezar (2005) established that "the [economic] neoliberal philosophy was one of the main forces driving the move away from the traditional charter between higher education and society, a tradition built on a communitarian philosophy of the public good" (p. 454). Economic neoliberals (a subset of those defined as perpetuating the era of conservative modernization described earlier [Apple, 2006]) are defined as people who believe in a free-market economy that includes institutions of higher education. Kezar's analysis shows the myriad ways that higher education for the public good has been reconceptualized as a privatized public institution through an economic neoliberal philosophy. If the shift of higher education from public/social good to a private/economic good is not interrupted by leaders in the academy, then "even if we want to alter the social charter [between higher education and society], it may not be possible to revitalize lost areas of the public good" (p. 26). The public and the private are interconnected and need to be addressed by higher education leaders as such when crafting higher education's public agenda.

Refinancing Postsecondary Education to address Access and Equity

St. John (2006) echoes the crisis mentality as demonstrated by other scholars in the *Interconnected and Advocacy* frame and contends that "the challenge of improving high school preparation, access to higher education, and degree attainment is far more serious than previously portrayed in official statistical reports on the subject" (p. xxi). He focuses on the role of finances in promoting educational access and equity in K-12 and postsecondary education. He uses John Rawl's theory of justice to reconceptualize educational reform as connected to justice and equity. In particular, St. John (2003) developed the balanced access model to recognize the importance of considering academic preparation, school and postsecondary finance strategies, and public policy simultaneously to understand the complexities of improving student pathways through K-16 systems.

St. John urges for a change in the current patterns of finance in order to "expand opportunity for college-qualified students from low income families" (p. 173). Specifically, he distinguishes between academic *and* financial access by thus defining these two distinct terms:

> *Academic access* is determined by institutional admissions decisions, which are based on a review of students' academic qualifications and applications for admissions. *Financial access* is the ability to afford initial and continuous enrollment; it can be influenced by governmental and institutional aid subsidies, college costs, and family incomes and savings. (Italics in original, p. 153)

By addressing both academic and financial access, St. John reconceptualizes educational access through a frame of social justice that directly addresses historical and contemporary issues of racism, sexism, and classism.

In *Education and the Public Interest: School Reform, Public Finance, and Access to Higher Education*, St. John (2006) extends his first analysis by introducing and testing a new framework for assessing the impact of public policies on student outcomes. Specifically, he examines state indicators of academic and financial access and offers a reanalysis of the National Education Longitudinal Study (NELS) regarding access to math classes, enrollment, and attainment. St. John contends that educational opportunity gaps widened as a result of academic reforms. The reforms have had a complex influence, where the policies have had a positive impact on test scores, a negative impact on high school graduation, and a positive influence on whether you will attend college upon graduation. St. John argues that these outcomes are far from equitable across race, gender, and class.

St. John also introduces the concept of educational citizenship that every citizen has the right to "secure employment sufficient to support families and to contribute to society" (p. 236). He argues for access to educational citizenship that will make a change toward social justice and equity in contemporary society and urges that current policymakers must address these inequities. This approach echoes the countless calls for connection between communities and universities (see the *Handbook of Engaged Scholarship* by Fitzgerald, Zimmerman, Burack, & Siefer, 2010). In order to address educational citizenship through policy, the concept of public interest must be redefined to include equal access to quality education for *all* members of society within the changing global context.

The Political, the Marketplace, and Educating for a Diverse Democracy

In *Take Back Higher Education: Race, Youth, and the Crisis of Democracy in the Post-Civil Rights Era* (2004), Henry Giroux, professor of secondary education, and Susan Searls Giroux, teacher of English and education, share their concern that higher education is under attack by economic neoliberals. The authors view public education as being redefined as a private good in order to further stratify the white upper/upper-middle class and the poor/working class who are predominantly people of color. The "take back" argument centers on the belief that higher education administrators and faculty members are passively enabling corporations to take over colleges and universities. Giroux and Giroux (2004) state that strengthening the relationship between higher education and society requires "rejecting the model of the separation between the public and the private domains and recognizing instead their mutual dependence" (p. 40). The lines between the historically separated entities are blurred and a university that acts as a public good must recognize that the false dichotomy of public and private has historically perpetuated racism and classism; a university might "open up a space for more than just a democracy" (p. 213). In this conceptualization, the public and the private are both critical to the success of a democracy. The authors further assert that the current, significant concentration on market values by higher education leaders stratifies the "haves" and the "have-nots" thereby decreasing the value of higher education as a public good. This argument is echoed by a number of other education scholars (Apple, 2006; Bowen & Bok, 1998; Brint & Karabel, 1989; Green & Trent, 2005; Hagedorn & Tierney, 2002; Labaree, 1997, 2007).

The Girouxs (2004) address how to revitalize the relationships between higher education and society through the creation of a "paideia" (p. 37), or a "prodigious political educational process," in order to develop the skills and abilities that an educated citizenry, participants in a diverse democracy, requires. In this frame, and similar to Pitkin and Shumer (1982), neither privileged nor oppressed people are excluded from access to higher education or participation in a democracy. In this vision of higher education, race and civic education cannot be extracted from the context of US multiracial and multiethnic society. This also means educating the entire public through multicultural education.

Finally, the authors in this *Interconnected and Advocacy* frame often refer to political theory and stress the importance of race, ethnicity, gender, and socioeconomic status to the relationships between higher education and society. The inclusion of all things political is reflective of the notions of social capital as defined by French sociologist Pierre Bourdieu. Bourdieu's (1986) definition of social capital connects three sources of capital—namely, economic, cultural, and social—in order to create an aggregate of resources linked to a network of relationships. Bourdieu defines social capital as grounded in theories of symbolic power and social reproduction where social capital is a tool of reproduction for the privileged. In this sense, social, cultural, and economic capital continues to be reproduced in a way that supports the privileged and marginalizes the oppressed. This is quite distinct from Coleman's (1988; 1992) use of social capital described in the *Balanced* frame.

AN ANALYSIS OF THE RELATIONSHIPS AMONG FRAMES

What are the relationships—or lack thereof—between these various perspectives of higher education and society? And, more importantly, what are the implications of these relationships? In this section I share four visions that further the conversation about the relationships between higher education and society by sharing general observations and implications about the connections between frames.

A More Comprehensive Vision of the Benefits of Higher Education for Society

The first general observation is that numerous scholars cite the Array of Benefits IHEP (1998) figure when talking about the benefits of higher education for society as it is clear and easy to understand. The Array of Benefits represents only three of the four frames that emerged from this literature review. For example, the *Private Frame* is found within the private/individual and private/social quadrants; the *Public* is found within the public/social quadrant; and the *Balanced* frame is completely represented. The *Interconnected and Advocacy* conceptualization is not currently represented in the model. If the model were more reflective of all four frames, addressed educational inequities, and discussed the fluidity between quadrants, then it could be utilized to advance relationships between leaders from all four different frames and not remain limited to describing only three cognitive processing models.

I have adapted the Array of Benefits model to include all four frames (See figure 2.2). *The Benefits of the Relationships between*

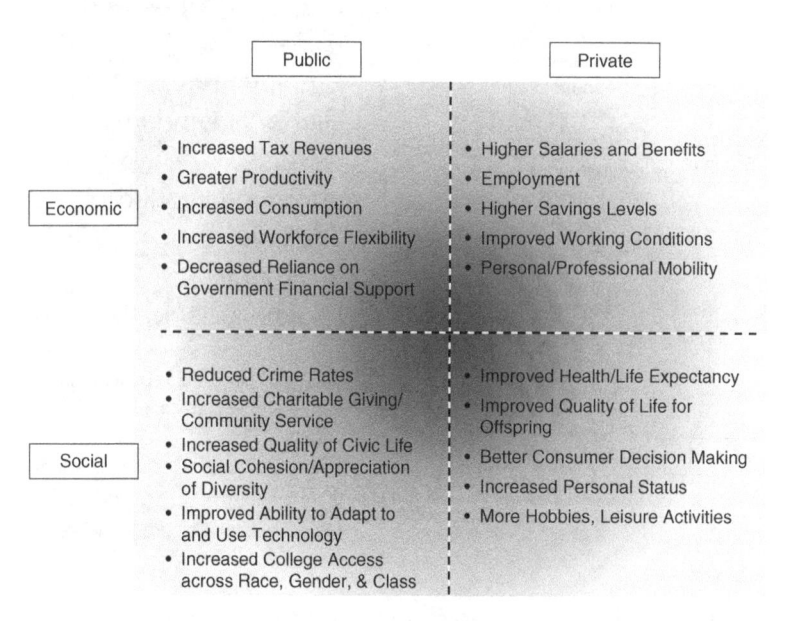

Figure 2.2   The Benefits of the Relationships between Higher Education and Society

Adapted by Penny A. Pasque from the Institute for Higher Education Policy (1998) in *Reaping the Benefits Defining the Public and Private Value of Going to College*[2]

Higher Education and Society model includes the addition of "Increased College Access across Race, Gender, and Class" as a public/social good as discussed in the *Interconnected and Advocacy* frame. The revised model also consists of dotted lines and a fading of color from the center (dark) to the edges (light) in order to portray the interconnections between quadrants. In a similar fashion to the Johary Window's[1] (Luft, 1970), the different quadrants expand and contract to reflect various relationships between the quadrants (See figure 2.3 & 2.4). In this model, the quadrants may take different shapes in order to reflect higher education's changing relationship with society at a particular point in time, or to represent the different frames of leaders. For example, in the *Advocacy* frame, the authors argue that the private/economic quadrant has recently been expanding, therefore, the private/economic quadrant is larger and its color

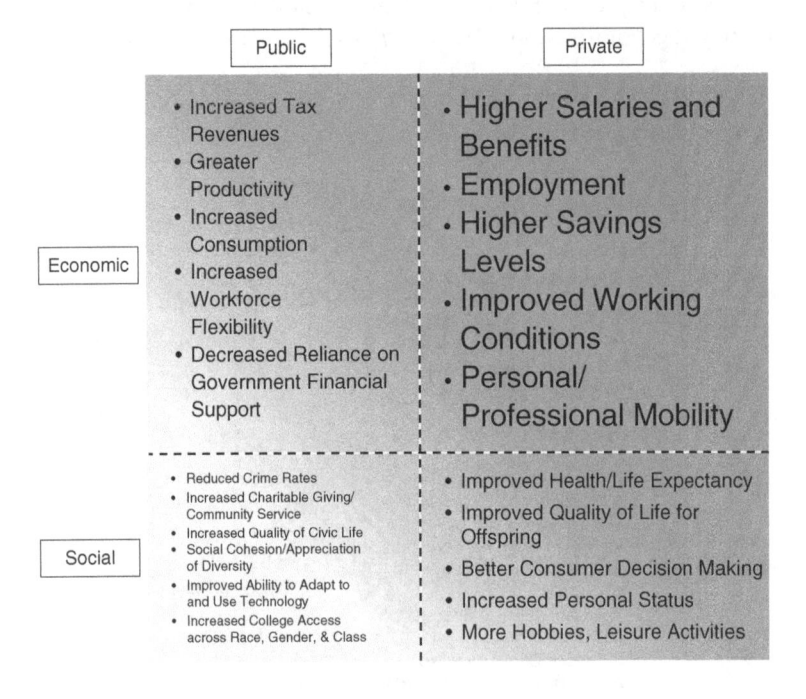

Figure 2.3 The Current Benefits of Higher Education for Society: A Private Good Frame

Adapted by Penny A. Pasque from the Institute for Higher Education Policy (1998) in *Reaping the Benefits Defining the Public and Private Value of Going to College.*

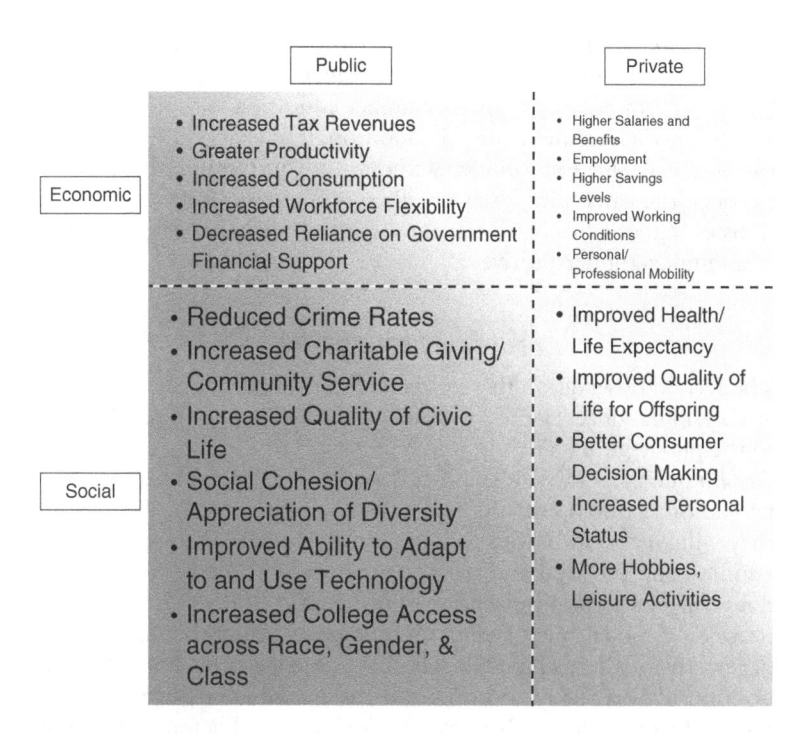

Figure 2.4 The Envisioned Benefits of Higher Education for Society: An Interconnected Frame

Adapted by Penny A. Pasque from the Institute for Higher Education Policy (1998) in *Reaping the Benefits Defining the Public and Private Value of Going to College.*

darker (See figure 2.3). Yet, as previously described, the scholars prefer a larger public/social quadrant where the public/social quadrant is larger and darker (See figure 2.4). This adapted model visually represents the fluidity between the quadrants.

If the *Interconnected and Advocacy* frame is not represented within a model of higher education benefits and the model remains limited to three conceptualizations, then the current model might further miscommunication among scholars with different frames and from different areas within the field. A comprehensive model that is both clear and easy, such as the adapted *Benefits of the Relationships between Higher Education and Society* model, could expand the conversations in order to increase understanding of the benefits of higher education to legislators, business leaders, and the public, a goal that is encouraged by McMahon (2009) in *Higher Learning, Greater Good: The*

Private and Social Benefits of Higher Education. This revised model has the potential to further the current literature and to help garner legislative and public support for higher education without eliminating issues of access and equity. In addition, if scholars continue to cite the Array of Benefits model without fully exploring the fluidity between the categories, then it will continue to limit conceptualizations of higher education's relationships with society to the exclusion of an important perspective.

Social Justice and Equity

The second observation that needs to be made explicit is that *all four of the frames* address issues of social justice and equity in relationship to the public good, albeit in different ways. For example, the *Private Good* frame takes the stand that economic disparities exist and that if people (undefined) are educated through postsecondary education, they will be able to become economically viable for themselves, their families, and the local and state community. The market perspective does deliver a certain conception of justice and equity. This cognitive processing model mirrors the "pull yourself up by the bootstraps" perspective—where all people who have an education should be in a position to financially contribute to the public good through their own individual success. It is crucial to note that this frame does not always specify *how* all people in US society will be able to gain access to higher education, *who* has yet to gain this access, or even that inequities across race, gender, and nationality exist in the current system in addition to class-based inequities. It also does not address other ways of contributing to society beyond individuals managing financially for themselves and their families. Solely a *Private Good* market system of higher education for the public good has the potential to further stratify the system of higher education in terms of race, gender, nationality, and class.

In the *Public Good* frame, inequities across race, gender, nationality, and class are addressed, where justice and equity may be achieved (or at least strived for) through educating students to understand the multiple causes of educational inequities, to commit themselves to work for a more equitable society and access to education, and to participate fully in a diversity democracy. It is through multicultural education and civic participation that all students (students with various social identities, including and beyond the identities mentioned here) may learn about inequalities, diversity, and the need for intergroup collaborations to work toward creating a just and equitable society. The concept of community, specifically community-university partnerships, is a foundational element in this frame. This frame does not focus on market influences or on how individuals can contribute to the public good by individually being economically successful.

The *Balanced* frame emphasizes increased understanding of both the public and private benefits of higher education in the creation of public policy. This conceptualization argues that a private frame is limiting in that it couches all higher education benefits as financial, or that when social benefits are named, they are simply "spillovers" and not given the emphasis they deserve. In this way, the focus of a balanced perspective is on recognizing the important economic benefits of higher education and increasing awareness of the multiple social benefits that incur when engaged in higher education. This framework does not always offer an explicit approach to issues of social identity, crisis, access, and in/equities or how these issues are connected to each other.

The *Interconnected and Advocacy* frame specifically names justice and equity as crucial to strengthening the relationships between higher education and society. "Equity" is intentionally used instead of "equality." From this perspective, equality may not be attainable until equity is actualized. Justice and equity across race, gender, nationality, and class are considered simultaneously through the interconnections of the public and private goods. Not only does this frame specifically name justice and equity as important, scholars also urge higher education leaders to change the current system of admission and financial support for students in order to educate more historically and contemporarily disenfranchised people in our communities. This argument does not foster fears of "the other," fears that may, in turn, reinforce protectionist tendencies and values of social reproduction, an argument commonly found in current higher education public agenda discourse (Gildersleeve et al., in press). Instead, this *Advocacy* framework creates awareness of reductionist notions of diverse populations and emphasizes the importance of educating students about inequalities, the need for collaboration across communities to achieve social change, and the significance of experience with diversity for civic engagement. It is the specificity of naming social justice and equity "for whom" (inclusive of all people) and "in what ways may this be possible" (for equity across race, class, nationality, gender, and multiple social identities) that distinguishes this frame's concepts about social justice and advocacy.

A Conflict of Visions of the Relationships among Higher Education and Society

The third observation is that if scholars continue to disagree among themselves about the definition and benefits of higher education for society, let alone the evidence and strategies for change in the system, then how can the system of higher education advance theory-to-practice through policy and campus initiatives toward change? Public policy scholar and historian Sowell helps to illuminate this point.

In *A Conflict of Visions: Ideological Origins of Political Struggles*, Sowell (2002) describes his sense of the public and private good and conjectures that the private good represents a "constrained vision" and the public good represents an "unconstrained vision" of society. In the constrained vision, individual benefits might persuade a person to act in hir[3] own self-interest or for "inner needs" that s/he "would not do for the good of his fellow man" (p. 22). Sowell believes that the way to combat the self-centeredness of benefiting the private, or individual good, is through intervention with trade-offs, an intervention that assists in balancing the public benefits to society. This vision is most fully captured in the *Private* benefit frame discussed above. Sowell's unconstrained vision is one where people are "capable of creating social benefits" (p. 23) without the need for the intervention of trade-offs.[4] Here, members of society follow through on a solution such as civic engagement because it is the right thing to do for the good of society. This vision is most compatible with the *Public Good* and *Interconnected and Advocacy* frames about higher education and society.

In addition to juxtaposing the constrained and the unconstrained visions of community members, Sowell positions the public good perspective in opposition to private good. He does not formulate a continuum because,

> in one sense [a continuum would] be more appropriate to refer to less constrained visions and more constrained visions instead of the dichotomy used here. However, the dichotomy is not only more convenient but also captures an important distinction... Every vision, by definition, leaves something out—indeed, leaves most things out. The dichotomy between constrained and unconstrained visions is based on whether or not inherent limitations of man are among the key elements included in each vision. The dichotomy is justified in yet another sense. These different ways of conceiving man and the world lead not merely to different conclusions but to sharply divergent, often diametrically opposed, conclusions on issues ranging from justice to war. There are not merely differences of visions but conflicts of visions. (p. 33)

In Sowell's dichotomy, viewing higher education as a public good negates the possibility of considering higher education also as a private good, and vice versa. People will view higher education only as an unconstrained vision *or* as a constrained vision. Sowell's vision includes no way for people with different conceptions of what higher education contributes to society to come to understand one another or reach a set of solutions for which they could work together. It follows that those who perceive higher education as a public good are diametrically opposed to those who perceive it as a private good, and recommendations for change to increase higher education's role in serving the public will not overlap. While Sowell's overall argument does not necessarily view the two as hopelessly dichotomous, he makes explicit that he prefers the private good, or constrained vision, as he argues for the "convenient" dichotomy.

Sowell's argument is supported by the Marwell and Ames (1981) free-riding experiment about investment in the private versus public good. Free-riding "refers to the absence of contribution towards the provision of a public good by an individual, even though he or she will not be excluded from benefiting from that good" (p. 296). The authors found that participants voluntarily contributed 40–60 percent of their resources to the public good despite the experiments' attempts to maximize self-interest. However, the authors also found that graduate students majoring in economics contributed an average of only 20 percent of their resources to the public good. The students studying economics were more likely to "free ride" than any other group of students. The students seem to be involved in a cycle in which they learn about the self-interest economic model and then apply them and perpetuate them in their own ways of thinking about the public good. Frank (2005) argues that students in economics should be exposed to both constrained and unconstrained notions of the public good in order to break this cycle of self-interest for economic students.

If Sowell's vision is taken at face value, then the primary implication is that the *Private Good* and *Public Good* frames cannot be integrated, as in the *Interconnected and Advocacy* frame, because private and public are diametrically opposed. The best Sowell would hope for are tradeoffs between the public and private. In my view, Sowell's vision (one implicitly or explicitly held by many voices in public discourse) will stifle progress toward strengthening the relationships between higher education and society. Moreover, scholars will not be able to see or communicate from "the other" vision but instead will continue communicating solely from their own perspective without fully comprehending or without listening to other perspectives.

Students and the public will hear competing information about higher education from both visions but may not be able to grasp a possible integration or decipher the full meaning of both perspectives; instead they may view the "other" perspective simply as propaganda. This could hinder an increase in awareness and understanding among leaders and slow down initiatives for developing greater public and legislative support for public colleges and universities, a concern already articulated by proponents of the *Balanced* perspective.

This thesis takes the position that the four frames that emerged from the literature are not mutually exclusive. The potential for advancing the conversation between people who implicitly (or explicitly and thoughtfully) follow either the constrained or unconstrained visions of Sowell may not be as bleak as Sowell's "convenient" dichotomy suggests. Similar to a pendulum, the extremes may, in fact, push the national policy conversation further than Sowell imagines. We, higher education leaders, need the extremes to push the conversation beyond the status quo. In this manner, each frame adds to the theory and practice of higher education for the public good and helps to increase awareness about various perspectives and benefits of the relationships between higher education and society. Yet, it is often the less radical perspectives (i.e., *Public* and *Balanced*) that receive the support needed for implementation, as they may not be perceived as extreme and as threatening as the other perspectives. It is important yet difficult to achieve discussion of all four frames among higher education leaders.

Higher education researcher Powell and sociologist Clemens (1998) also believe that the public good will always be "unsettled and contested and is part of the unsettled and contested nature of politics itself" (p. 4). The argument will never come to a conclusion. It is through conflict, dissention, and discussion that ideas are furthered. St. John, Kline, and Asker (2001) assert that arguments about higher education increasing human capital, causing economic development, and affecting equity issues enable liberal interests (i.e., affirmative action and educational equity for social justice) and conservative interests (i.e., economic development) to come together.

Visions of Capital Further Miscommunication

A fourth observation is that some cognitive processing models are reminiscent of each other, yet the principles of the conceptualization may be very different, such as with the *Public* and the *Interconnected and Advocacy* concepts of the relationships between higher education and society. Leaders might think that they are supporting one another, or that they agree with each other based on a similarity in language, and yet might actually labor against one another. For example, a number of scholars from each conceptualization mention "capital," yet each defines it with extremely different theoretical underpinnings.

The *Private Good* conceptualization addresses "human capital" as a primary component in the relationship between higher education and society. Human capital is highly connected to individual wage rates, state and individual rates of return on investment, and national and local economic growth (Becker, 1964/1993; Blinder & Weiss, 1976; Gottlieb & Fogarty, 2003; Weiss, 1995). This is distinctly different from how the other three frames utilize "social capital." The *Public Good* conceptualization addresses social capital as imperative to serving the public good and the scholars often cite Putman (1995; 2001) for a definition. Putnam (1995; 2001) defines social capital as the value of social networks. The *Balanced* and *Interconnected and Advocacy* frames, however, address the importance of human and social capital. The *Balanced* concept addresses both forms of capital as quantifiable, as is found in Coleman's (1988) definition of social capital that is utilized to determine a quantitative relationship between social capital and educational outcomes. The *Interconnected and Advocacy* frame addresses social capital from Bourdieu's (1986) perspective, where social, cultural, and economic networks facilitate access to social capital. Bourdieu's construction of social capital also includes issues of political capital and systemic oppression. Further, Hagedorn and Tierney (2002) as well as Giroux and Giroux (2004) specifically name Bourdieu and extend Bourdieu's definition to include issues of cultural capital (also see Tierney, 2003).

If the relationship between frames and the notion of "capital" remains unexplored as related to the relationships between higher education and society, then this ambiguity could add to a further disconnect between perspectives and miscommunication between leaders. Scholars might assume they mean the same thing when they use "social capital," yet the words often have vastly different definitions. Lack of mutual understanding has important consequences when trying to work with legislators, the public, or colleagues at the same institution, and when attempting to garner support for policies to increase access to higher education for educational equity.

Inclusive and Exclusive Visions of "The Public"

A final observation is that each scholar assumes that a "public" exists in US society. Among authors addressing the relationship between higher education and the public, I found none argued that the "public" does not exist. In addition, who is included and who is excluded (as subject) in the public is different for each author. For example, in the *Private Good* frame, a number of scholars talk about the public as an abstraction—without defining who is included or excluded in their concepts of the public or society. Very few scholars mention class, gender, and race. And, as Urrieta (2008) reminds us, not all communities may share the same visions of the public good.

By failing to define the public, there exists an assumption that the public, as subject, follows the dominant paradigm and includes people who have traditionally been considered in the "public" such as able-bodied, white, Christian, heterosexual men and those who assimilated to US culture (Bell, 1997). The *Public Good* frame includes every participant in the diverse democracy as part of the public, regardless of citizenship status (e.g., legal permanent resident, undocumented immigrant). This definition is open-ended and inclusive. Some of the *Balanced* scholars define the public as diverse and inclusive (Baum & Payea, 2004; IHEP, 1998; McMahon, 2009; Wagner, 2004) and some do not (Boulus, 2003; Callan & Finney, 2002; Public Sector Consultants, 2003). The *Advocacy* scholars each include the diverse public in their definition and specifically name the underrepresented and disenfranchized as people to include in definitions of the public.

If the public, as subject, is not defined as we try to strengthen the relationships between higher education and society, then an assumption about whom the term "public" includes and excludes exists and by default the definition will include only people with dominant social identities. This could lead to miscommunication and misinterpretation between conceptualizations if the definition of the "public" is left to the readers' interpretation. Furthermore, if the subject/public is intentionally defined as inclusive, it enables scholars with each frame to address the pervasive historic inequities in order to strengthen the relationships between higher education and all of society. Lack of attention to intersectionality—the intersections of class, race, ethnicity, gender, sexual orientation, ability, and various social identities—creates serious limitations in college contexts (Chesler, Peet, & Sevig, 2003; Myers et al., 1991; Sevig et al., 2000). If the public is not defined as inclusive, then people historically excluded from that public will not be included as the future of higher education is envisioned, nor will they reap the benefits identified in any of these frames. In this sense, the "subject" becomes more of an "object" to discard or ignore.

Summative Statement and Next Steps

Each of the four frames—or cognitive processing models—offered in this chapter proposes a different set of ideas about how to make change for the public good, where the "public" is inclusive of all people in the United States. Although I hold my own cognitive processing models, in this chapter I argue that knowing more about multiple frames expands our awareness of the varying perspectives held by higher education leaders. Acknowledging multiple existing frames also helps us explore the tensions between perspectives in order to make educated choices and break the cycle of what is currently not working in the system of education.

As Miller and Fox (2004) point out, this macro-analysis is not enough. The perspectives of higher education leaders and their cognitive processing models are often expressed in and through face-to-face discourse. As a bridging micro-analysis, I chose to explore a national conference series where higher education leaders and policymakers came together to discuss strengthening the relationships between higher education and society. I asked the following orienting research question: "What happens when higher education leaders come together to discuss how to strengthen the relationships between higher education and society?"

I decided to use an emic approach where the findings emerge from the participants, rather than an etic approach where a framework (such as these four frames) is imposed on the conference transcripts. In this way, I hoped to not limit the larger study to the four frames found in this literature review but to continue to follow this emergent approach in order to uncover main concepts that emanate from the discourse of the leaders themselves. This process also allows me to explore the similarities and differences between what the higher education leaders write about and what they speak about in a national policy context. I discuss the similarities between the literature review and face-to-face discourse in subsequent chapters.

CHAPTER 3

Research Design

Lather (2003) counters the tendency for theoretical determinism in critical scholarship in her discussion of issues of validity in qualitative research. Yet, choosing a single theory for a research study, even when using critical theory, still exists as the primary approach in higher education research. As Abes (2009) encourages,

> the researcher should consider experimenting with the choice and application of theoretical perspectives bringing together multiple and even seemingly conflicting theoretical perspectives to uncover new ways of understanding the data. Rather than being paralyzed by theoretical limitations or confined by rigid ideological allegiances, interdisciplinary experimentation of this nature can lead to rich new research results and possibilities. (p. 141)

In this chapter, I offer theoretical and methodological approaches that resist theoretical determinism in education research and search for new possibilities reflective of critical inquiry's emancipatory and empowering objectives (Brown & Strega, 2005; Kincheloe & McLaren, 2005). The complexities of the approach described here strengthen its trustworthiness (or validity, goodness, congruence—to be discussed in more depth later in the chapter) as it combines multiple lenses through which to explore the discourse. This approach emerged from an iterative research process of reading, writing, reflecting, and dialoging with research colleagues over the course of many years and is not intended to be static.

After describing the theoretical and methodological framework, I discuss the elements of trustworthiness in this qualitative research study, including triangulation, member checking, and researcher reflexivity. Finally, I describe the specific research methods that were used in the three major aspects of this study: (1) the macro-analysis of the literature; (2) the micro-analysis of the discourse through a post-positivist approach, which helped me locate the phenomena and study it in more detail; and (3) the micro-analysis of the discourse through a critical approach, which provides added depth to this analysis and discussion. The intentional and simultaneous consideration of the macro- and micro- language around this topic allows for a more complex analysis of the intricacies of the discourse by leaders in higher education (Miller & Fox, 2004).

THEORETICAL AND METHODOLOGICAL APPROACH

This theoretical approach focuses on the intersections of (1) discourse methodologies (e.g., critical discourse analysis, narrative analysis, conversation analysis) and (2) social identity theoretical lenses of race, ethnicity, gender, sexual orientation, and class (e.g., critical race theory, feminist theory, queer theory, and post-Marxist theory). For clarity, I have drawn the connections between these theories in figure 3.1.

As figure 3.1 portrays, I take an overall critical inquiry approach to this study while using the connections between discourse methodologies and social identity theories as a lens, each of which is explained further in the next sections. At the point of intersection of these approaches, the Conceptual Model of Multiple Dimensions of Identity by Jones and McEwen (2000) is particularly useful. This model reflects the multiple social identities of an individual (age, class, gender, race, religion, sexual orientation) where each approach is inextricably linked to the other and cannot be separated in real-life contexts. For example, I am a gendered, raced, classed, and sexually oriented being all at the same time (among other identities). This fluid conceptual model is informed by identity-based theories such as feminist theory, critical race theory, queer theory, and class-based theory—each of which highlights a particular element of a person's identity and yet is often interwoven with (1) other social identities and (2) historical, cultural, and sociopolitical contexts. Each of these identity theories requires scholars to consider an issue or problem in relation to gender, race, sexual orientation, ethnicity, socioeconomic status, and class, among others, and inherently leads me to consider leaders' verbal and written discourse through multiple lenses as connected to systemic issues of sexism, racism, heterosexism, classism, and other isms.

It is important to note that the Conceptual Model of Multiple Dimensions of Identity was developed from the individual within a context and, for the purposes of this study, is to be viewed with the individual, institutional, *and* structural milieu in mind (Pincus, 2000). This expansion, inclusive of and beyond the representation of

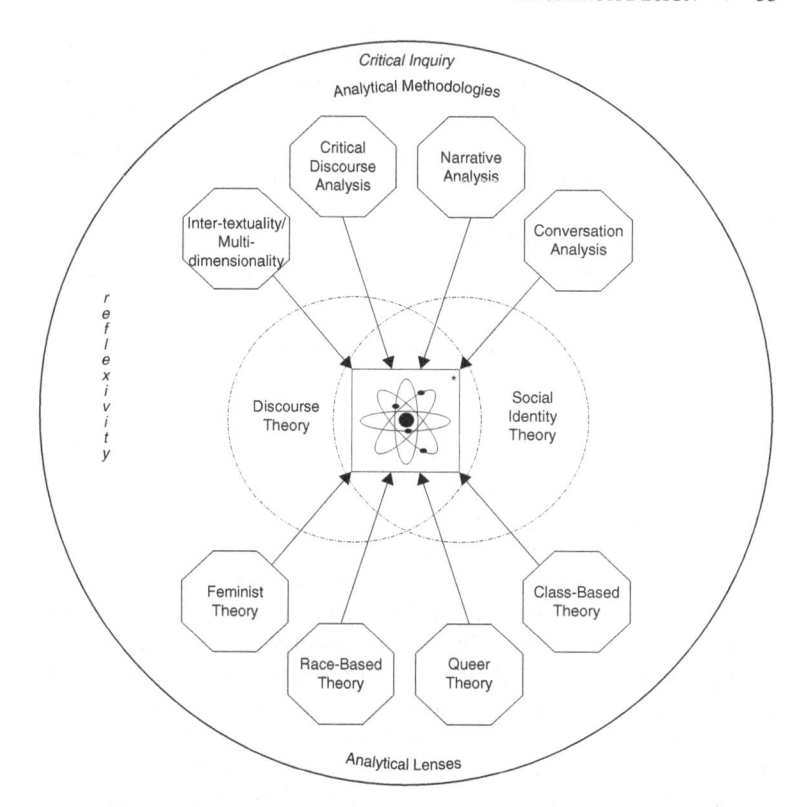

Figure 3.1 Theoretical Lenses for Locating the Phenomena

*A Conceptual Model of Multiple Dimensions of Identity by S. Jones & M.K. McEwen (2002) is embedded within this figure. For the purposes of this theoretical approach, the model (originally developed for the individual) signifies the individual, institutional, *and* cultural context (see Pincus, 2000).

the individual, is vital in order to encapsulate the complex and inter-connected nature of individual identity, social capital (Bourdieu, 1986), and myriad levels of power (Foucault, 1976). I describe each aspect of this theoretical approach in more detail in the sections below and then describe my rationale for using this complex approach, which intentionally connects social identity and discourse.

Discourse Methodologies

There are two discourse methodologies central to this analysis: critical discourse analysis and narrative analysis. Multidimensionality and conversation analyses—also used to analyze the discourse but in a less central manner—are described in later sections in order to help provide additional context. These discourse methodologies are the tools I employ to explore the cognitive processing models of higher education leaders' verbal discussion about higher education's relationships with society. These tools provide a way to explore what is said (content), how it is said (process), and how people respond to (or do not respond to) what is said (content and process) (Trenholm & Jensen, 1992).

Critical Discourse Analysis

Critical discourse analysis (CDA) may be understood through a description of the binary sociopolitical processes of voluntarism and determinism (Erickson, 2004). Specifically, Erickson describes "voluntarism" as the perspective from which the individual is ontologically fundamental, wherein people believe that "whatever happens in society is ultimately the result of the exercise of individual effort and will" (p. 112). In this definition of voluntarism, the individual has control over the local and global processes around him/her/hir. More pointedly, individuals may make change in their environment through their own effort and will. Alternatively, Erickson describes "social determinism" as the mutually exclusive and polar opposite of voluntarism, where "society, as a whole and in its parts, is ontologically (and empirically) prior to the individual" (p. 113). In this definition of social determinism, society dictates the fate of an individual. More specifically, a socially deterministic society is one where hegemony and systemic oppression prevails at the expense of the individual. Such a dichotomy is reflective of the public and private binary described more thoroughly in chapter 2.

CDA rejects the dichotomy between voluntarism and determinism and sees the two as interconnected. In this interconnected approach, human agency is the factor that works within and between voluntarism and determinism. Explicitly, human agency helps individuals make change in society and simultaneously influences society to make changes that affect individuals. Through a critical discourse analysis theoretical lens, principles of voluntarism and determinism need to be viewed concurrently and as interdependent.

Erickson's rejection of the binary categories of voluntarism and determinism correlate with the multiple theoretical lenses of identity. In Erickson's case, the individual identity is not void of the sociopolitical context, and vice versa. Similarly, theoretical lenses of social identity are not void of the historical, cultural, and sociopolitical contexts. Further, critical discourse analyses and the theoretical lenses of social identity both include the concept of human agency and the connection between the individual and society. In this context, the theoretical lenses of critical discourse analysis and social identity are to be included in concert with each other as I explore the narratives of higher education policy leaders and their perceptions of critical issues and the public good.

Fairclough (2001) extends the definition of CDA and describes it as a way of studying "how language figures in social processes" (p. 229). With this connection between language and social processes come areas within which social life is produced—such as social identities, means of production, economics, cultural values, political and social relationships, and consciousness. In the context of this study, I explore the ways in which language about the relationships between higher education and society—including language about social identities and the social identities themselves—figures in social processes in the context of higher education. Fairclough states that CDA is *critical* because it

> seeks to discern connections between language and other elements in social life which are often opaque. These include: how language figures within social relations of power and domination; how language works ideologically; the negotiation of personal and social identities (pervasively problematized through changes in social life) in its linguistic and semiotic aspect. Second, it is critical in the sense that it is committed to progressive social change; it has an emancipatory "knowledge interest" (Habermas, 1971). (p. 230)

Specifically, CDA relates to this particular micro-analysis of the national policy conference series in that it explores multiple dimensions of discourse and simultaneously acknowledges power and domination, considers the ideological perspectives of leaders, and explores the complexities about personal and social identities in a sociopolitical context. The higher education leaders who have content knowledge about and power in the system of higher education are the focal point of this analysis. These leaders often talk about making change in the system in order to strengthen the relationships between higher education and society; however, a close, critical analysis of the ways that power operates in social relations, the ideologies or frames of the leaders, and the ways that social identities of the leaders (or social identities of people who feel the implications of leaders' talk and action) are formed is not often considered in the existing literature.

As Cameron (2001) explains, CDA is concerned with the "hidden agenda" of discourse or its "ideological dimension" (p. 123) where choices about discourse are not viewed as random but as ideologically patterned. This CDA concept connects to the pilot study analysis, which uncovered the patterned cognitive processing models of higher education leaders and the different ways in which each leader approached the conversation at hand (Pasque & Rex, 2010). This particular study also explored the ideological positions of higher education leaders by identifying emergent patterns in narratives on a larger and more comprehensive scale.

In addition, critical theory always tries to move beyond the theoretical realm to create concrete social change (Kincheloe & McLaren, 2005), and discourse analysis hopes to contribute to our understanding and solving of issues and problems (Gee, 2005). Critical theorists are "concerned in particular with issues of power and justice and the ways that the economy, matters of race, class, and gender, ideologies, discourses, education, religion, and other social institutions and culture dynamics interact to construct a social system" (Kincheloe & McLaren, 2000, p. 281). This critical discourse analysis approach adds an action-oriented lens and has the potential to shed new light on a situation not often discussed in or out of the academy—the power dynamics, identities, and ideologies of leaders who are interested in the topic of strengthening the relationships between higher education and society.

Narrative Analysis

The cultural practice of storytelling about personal experience through narratives constitutes a significant means through which people position themselves and are positioned (Walton, Weatherall, & Jackson, 2002). These discourses frame our understanding of an identity of the self or of something other than ourselves. In this case, the narratives portray how higher education leaders position themselves as well as each other and/or perceive higher education's relationships with society. The national conference series elicits such discourse from higher education policy leaders.

Daiute and Lightfoot (2004) define narrative analysis as "different ways of conceptualizing the storied nature of human development" (p. x). Narrative analysis,

> is appealing because its interpretive tools are designed to examine phenomena, issues, and people's lives holistically; is excellent contexts for

examining social histories that influence identity and development; generates unique insights into the range of multiple, intersecting forces that order and illuminate relations between self and society; and permits the incursion of value and evaluation into the research process. (p. xi–xiii)

This definition of narrative analysis draws strong parallels with the critical discourse analysis and the lenses of social identity and provides tools with which to consider participant worldviews from a holistic perspective. Narrative research also strives to question common understandings and, at times, evokes dissonance (Coulter & Smith, 2009). It is also important to note that significant research links narratives and identity development (See Sarbin, 2004).

In addition, oral and written narratives capture cultural models of human agency, schemas for relationships, goals, and subsequent actions to fulfill the goals (Lee, Rosenfeld, Mendenhall, Rivers, & Tynes, 2004). As Daiute (2004) states, narrative is a "medium of identity development, healing, learning and planning for the future" (p. 112). In this way, this study is not a critique for critique sake but focuses on utilizing the narratives from higher education leaders in a way that is instructive for social change and educational equity.

To add further rationale for the inclusion of narrative analysis, Bruner (1990) states that narratives serve to (1) highlight culturally recognizable explanations or interpretations, (2) attend to the knowledge and intent of listeners and the protagonists in their stories, (3) make use of a culturally commonsensical epistemology, and (4) take a moral or evaluative stance relative to the events in the story (also see Walton & Brewer, 2001). Further, as Coulter and Smith (2009) point out, this approach allows readers to make sense of the study in their own way where "multiple interpretations by multiple readers are expected and promoted" (p. 578). Taken together, narrative analysis serves as a useful lens to help further explore leaders' perspectives within a sociopolitical context as the focus is on the participants as speakers, the participants as listeners, *and* the readers.

Social Identity Theories

In addition to approaching this study from critical discourse and narrative methodological perspectives, I also draw from identity theories that focus on social identity (Bell, 1997; Hardiman & Jackson, 1997; Jones & McEwen, 2000; Tatum, 2001).[1] The connections between social identity and discourse are clearly encapsulated in CDA and narrative analysis. In addition, social identity theories are often used when researchers explore the discourse of participants in order to further understand a particular topic. I am interested in the areas of overlap between these two areas (as visually displayed in figure 3.1) since an approach that combines discourse theories and social identity theories will provide me with a rigorous lens through which to consider the following orienting research question: "What happens when higher education leaders come together to discuss how to strengthen the relationships between higher education and society?" With this complex lens, both the language and social identities of participants are privileged in this research context, as opposed to viewing the discourse of leaders through one lens in the absence of the other.

From the social identity approach, people with "target" identities are "members of social identity groups that are disenfranchised, exploited, and victimized in a variety of ways by the oppressor and the oppressor's system or institutions" (Hardiman & Jackson, 1997, p. 20). In contemporary US society, this includes people of different abilities, people of color, non-Christians, the working class/poor, gays/lesbians/bisexuals, and women/transgendered people. Agents include "members of dominant social groups privileged by birth or acquisition, who knowingly or unknowingly exploit and reap unfair advantage over members of target groups" (Hardiman & Jackson, 1997, p. 20). In contemporary US society, this includes people who are able-bodied, white, Christian, middle and upper-middle class, heterosexual men and those who have assimilated to US culture.

It is important to note that it is impossible to separate various personal social identities from social group memberships (Bell, 1997). Social identities are fundamentally connected to the cultures in which they are embedded (Bakhtin, 1981; Vygotsky, 1978). Identities are also fluid, they change as people move through life; the sociopolitical environment may also change over time. Theoretical lenses such as feminism (Astin & Leland, 1991; Bem, 1993; hooks, 1984/2000; 2000, Irigaray, 1974; Nussbaum, 1999, Pasque & Nicholson, in press; Tong, 2009), race theory (such as critical race theory) (Valdes et al., 2002; Crenshaw, 2002; Delgado & Stefancic, 2002 and also see Torres, 1998 and Yosso, 2005), queer theory (Abes & Kasch, 2007; Berzon, 1996; Blumenfeld & Raymond, 2000; Butler, 1990/1999; Herek, 2001; Morris, 2000), and class-based theory (Davis, 1998; Dews & Law, 1995; hooks, 2000; for post-Marxist theory see Lather, 2003) highlight one identity over others (such as gender, race, sexual orientation, and socioeconomic status respectively) and often address the interconnections within and between social identities. In addition, they often do in-depth exploration of ideologies, domination, and subordination in historical and sociopolitical contexts from a lens that privileges that particular identity.

I am certainly not the first scholar to draw upon more than one theoretical lens at one time when exploring ideological perspectives. In her intellectual and activist work, Angela Davis (1998) connects race, class, and gender as she analyzes culture, capital, race, and gender in the United States. bell hooks (1984/2000; with Cornell West, 1991; 2000) addresses the intersections of gender, race, class, and sexual orientation while urging for critical reflection on contemporary issues in a historical and sociopolitical context. LatCrit, as defined by Daniel Solórzano and Dolores Delgado Bernal (2001), is a "theory that elucidates Latina/Latinos' multidimensional identities and can address the intersectionality of racism, sexism, classism, and other forms of oppression" (p. 312). Maurianne Adams, Lee Anne Bell, and Pat Griffin (1997) bring together twenty-two authors to address the intersections of racism, sexism, heterosexism, anti-Semitism, ableism, and classism in *Teaching for Diversity and Social Justice: A Sourcebook*. They continue this work with others (2000) in *Readings for Diversity and Social Justice: An Anthology on Racism, Antisemitism, Sexism, Heterosexism, Ableism, and Classism*. In addition, Carlos Alberto Torres (1998) critically discusses the challenges of feminism, critical race theory, and postcolonialism to his theory of citizenship when discussing democracy, education, and multiculturalism in education.

Moreover, the social identity lenses offer perspectives that underscore the experiences of people of a certain identity within a historical, cultural, and sociopolitical context. For example, the feminist lens considers the discourse in a manner that privileges women. Further, feminists approach feminism from extremely different perspectives (i.e., liberal, radical, Marxist, psychoanalytic, care-focused, womanist/Black, Chicana, multicultural, global, postcolonial, and eco feminism) and in each case, the ways in which women have been included and excluded are stressed. Many feminist perspectives actually incorporate the intersections of more than one social identity. Race-based theories (including critical race theory) emphasize the social construction of race and provide a lens with which to consider discourse that highlights race and racism. For example, critical race theory advocates that to deny race, ethnicity, and other social identities in qualitative research studies perpetuates a myth of equality (Parker, 1998). Queer theory emerged from feminist lenses and stresses lesbian, gay, bisexual, transgender, and heterosexist perspectives. Queer theory provides an LGBT-centered lens with which to consider the discourse. Finally, class-based theories (including post-Marxist theory) pull forward the individual, institutional, systemic, and historical ways in which socioeconomic status operates—and has operated—in US society. *Importantly, none of these identity-based theories focuses on one identity in isolation.*

As I approach this study, I draw from the social identity theoretical lenses of race, ethnicity, class, and gender and leave ability, age, sexual orientation, religion, and other social identities for future research studies. As Audre Lorde (1983) argues, there is no hierarchy of oppressions. One identity is not necessarily "better" or "worse" than another; it is not useful to argue that one identity is more oppressed than another as it does not necessarily break the cycle of oppression for any group. As such, through the use of the Conceptual Model of Multiple Dimensions of Identity (Jones & McEwen, 2000), I hope to address the interconnections between multiple participant identities and implications of participants' views held simultaneously at any particular point in time.

Connecting Discourse Methodologies and Social Identity Theories

For a researcher, theory and methodology are inextricably linked and must be congruent in order to promote research integrity (Jones, Torres, & Arminio, 2006). As mentioned above, critical discourse analysis and narrative analysis include elements of social identity and social identity theories are often used to explore discourse. A number of scholars specifically connect identity with discourse (Goffman, 1981; Hall & Bucholtz, 1995; Johnstone, 2002; Tannen, 1993; 1994). For example, performances of identity relate to how social identity (ability, age, class, ethnicity, gender, gender expression, race, religion, and sexual orientation) is interconnected with and cannot be separated from discourse. This humanistic perspective explores the ways in which verbal language is a representation or a performance of a person's social identity. By way of example, Tannen's (1993) research focuses on how women and men communicate their identity through the use of various linguistic strategies. Tannen explores language and topics such as power and solidarity, indirectness, interruption, silence, and conflict. In this way, Tannen explores performances of gender identity. In another example, Foucault's (as cited in Foucault & Pearson, 2001) historical research focuses on the contradictions between discourses and the ways in which the self is pulled in different directions by discourse. Anna Deveare Smith's (1994) performance ethnography research is another way in which social identity and discourse are connected. Deavere Smith states, "Words are not an end in themselves. They are a means to evoking the character of the person who spoke them" (p. 1).

By intentionally developing this complex and congruent theoretical and methodological approach to this study, I am rejecting the tendency toward theoretical determinism as argued by Lather (2003) and simultaneously attempting to further the educational scholarship that includes multiple lenses as urged by Abes (2009). What I gain is an innovative way in which to view the cognitive processing models of higher education leaders in national policy conversations, a way that acknowledges the interconnection between discourse and identity as they are not mutually exclusive but inform each other. As I view the discourse of leaders through this intentionally interconnected lens, I raise consciousness (my own and the readers') about my processes as a researcher, strengthening researcher reflexivity and trustworthiness (Jones, Torres, & Arminio, 2006).

Let me explain this concept in more detail using an example of my own reflexive writing developed during this research study. Many higher education administrators and faculty were, at one time, involved in Greek life, residence life, student government, or other co-curricular student activities. These organizations often have t-shirts made for their organization or for a specific event. When creating a picture on a t-shirt, each color is placed on the shirt using a "screen." As each screen (color) is placed on top of the previous one, more of the full-color picture can be seen. It is through the use of multiple screens that we are able to see an array of colors and various shading in the picture. Different combinations of screens enable a viewer to see different aspects of the picture and multiple screens are needed to complete a comprehensive picture. The same is true of research design; different combinations of theoretical and methodological approaches enable us to see the various aspects of the higher education picture.

I offer another example by way of another metaphor that also emerges from my own experience—in the theatre, various moods and feelings are often communicated through an array of colors (with various hues and values) of light. For example, a thin layer of plastic with a slight color may be placed in the front of a light fixture to alter the color that radiates from the light. This thin sheet of plastic media is called "gel." The piece of gel will change the color on stage from white to—for example—blue. If a yellow piece of gel is added to this light, the hue on stage turns from blue to green. However, this hue of green is different from that formed through a single green piece of gel; the green gel projects a different shade of green than the blue and yellow together. In addition, if three lights project from various places in front of the stage (one from stage left, one from stage right, and one from the footlights), one backlights the stage area, and each has a different combination of colors of gel, together the colors project a particular mood on stage. Through one configuration of lights and gels, the audience may see a fiery argument between two people; through another, they may see in the background more tension between various people that contributes to how they perceive the argument on center stage. Different combinations of lenses reveal various aspects of the same story.

To view discussions between higher education leaders through one theoretical perspective is important, but I argue that to use multiple perspectives at various times highlights different aspects of the conversations in an instructive manner. This approach provides a strong, inclusive lens through which to consider the discourse of higher education leaders. I realize that I may be sacrificing depth in order to include the intersections of multiple discourse methodologies and identity theories. Additional research on discourse between leaders is needed, and one option is to explore the conference series further through each particular theoretical lens and then to return to an analysis that connects the social identity theoretical lenses. Therefore, I offer a chapter in this book that addresses this issue: in chapter 6, after an important and overarching analysis in chapters 4 and 5, I offer a more in-depth analysis specifically utilizing feminist theoretical perspectives (Tong, 2009) and methodological approaches (Harding, 1987; Ramazanoğlu, 2002). In this way, I provide further depth of analysis and encourage additional depth of analysis utilizing different approaches in the future.

Trustworthiness, Validity, Quality, and Goodness

As Creswell (2003) points out, validity in qualitative research designs is quite different from that in quantitative research designs. "Valid" research has been variously described as validity, trustworthiness, quality, and goodness, to name a few. There are also debates in qualitative research in terms of what constitutes validity, trustworthiness, quality, goodness, and the politics of evidence (Denzin & Giardina, 2008). For example, Eisenhart and Howe (1992) describe five general standards for quality educational research: (1) fit throughout the research design, (2) effective application of data collection and analysis, (3) coherence of prior knowledge, (4) internal and external value constrains, and (5) comprehensiveness. Lincoln and Guba (1985) provide their use of trustworthiness as represented by truth value, applicability, consistency, and neutrality and argue against validity in qualitative research. Lincoln and Guba also articulate how the meaning of trustworthiness is different in each research study—for example, the nature of threats to trustworthiness and minimization of these threats will be different in each context.

As previously mentioned, Lather (2003) argues against theoretical determinism to intentionally increase validity. Lather states, "*Triangulation*, expanded on the psychometric definition of multiple measures to include multiple *data sources, methods*, and *theoretical schemes*, is critical in establishing data trustworthiness. It is essential that the research design seek counterpatterns as well as convergences if data are to be credible" (p. 191). Lather reconceptualizes validity in an attempt to articulate an approach to empirical research that both advances theory-building and empowers the participants in research studies, a tension that has taken center stage in very few researchers' concepts of validity. Lather asserts that the rigor and trustworthiness of openly ideological research processes can be established through a more "systemic approach to triangulation and reflexivity, a new emphasis for face validity, and inclusion of the new concern of catalytic validity" (p. 206) where catalytic validity departs from a positivistic approach and is the degree to which the research process reorients, focuses, and energizes participants in what Friere (1973) terms "conscientization," that is, knowing reality in order to better inform it.

Borrowing a term from Lincoln and Guba (2000), Jones, Torres, and Arminio (2006) argue for "goodness" in quality qualitative higher education research (Also see Arminio & Hultgren, 2002). Goodness is offered as new language for judging quality qualitative research, one that focuses on the congruence between epistemological, theoretical, and methodological lenses along with the methods of a study. Finally, Burke Johnson (1997) discusses the strategies used to promote validity and includes triangulation, participant feedback, and reflexivity (Also see Krefting, 1991).

In this research study, a number of different strategies are used to strengthen its trustworthiness. These include (1) a pilot study (Pasque & Rex, 2010), (2) triangulation of the pilot study, macro-analysis of the literature review, and micro-analysis of the current research study (3) member checking or participant feedback, (4) rich, thick description to convey the findings, and (5) reflexivity. I describe each below (see Pasque & Rex for the pilot study).

Triangulation

I was able to address trustworthiness through triangulation of pilot study, the macro-analysis (language in the literature review of articles, books, chapters, and speeches), and the micro-analysis (language of leaders at a national conference series) in various ways. Triangulation is a research strategy that involves the utilization of several different methods to consider the same data (Denzin, 1978). By exploring both the macro and micro policy discussions simultaneously, I am able to "construct bridges between different approaches to social life, particularly perspectives that focus on macro- and microscopic issues" (Miller & Fox, 2004, p. 35).

Specifically, I triangulated the findings from three different sources of information, each of which emerged independently: the pilot study, literature review, and the larger conference series study. I was able to view the language shared by higher education leaders with different theoretical frames using different methodologies over a long period of time. This process for analysis added to the complexities of the findings. In each instance, participants offered similar cognitive processing models, or frames, as found through the emergent thematic, discourse and narrative analysis; some of the same issues emerged over and over again until a saturation point was reached.

The second way in which I employed triangulation techniques was through sharing the findings of the pilot study with a research team on a weekly basis over the course of four months. The research team reviewed the findings, provided feedback, and posed questions for further consideration. Third, I shared the research findings and/or theoretical framework for the current study with research colleagues and committee members at different points in time in order to elicit thoughtful feedback. Finally, Drs. Lesley Rex and Ed St. John served as co-analysts for this study. They consistently reviewed my findings, argument, and description of the research process and provided useful feedback and questions for discussion.

Triangulation in future studies could additionally include the ways in which emergent themes in the literature review are represented (through an etic or imposed theoretical framework) in the discourse during the conference series. For this type of analysis, I would take the frames that emerged from the literature review and uncover the ways in which these frames were or were not represented in the conference series (Johnstone, 2002). Analyses across race, gender, age, position, and other social identities and roles may also be beneficial; one analysis across gender is offered in this volume.

Member Checking

Member checking, or taking the findings back to participants for review, was an important element that added to the trustworthiness of this study (Jones, Torres, & Arminio, 2006). I chose four participants who represented different social identities and roles during the conference series. No invited member checker declined to participate. The known member checker identities include two women, two men, a Latina/o, an African American, two white people, a graduate student, a postdoctoral researcher (currently an assistant professor), an assistant professor, a full professor, and an organizer. I intentionally do not connect the gender, race, and role of the member checkers in order to ensure confidentiality.

Each member checker was provided with a later draft of the findings, asked to read the draft, and then meet to discuss their thoughts. Each meeting lasted more than 1 hour and 15 minutes and under 4 hours. The meetings were not audio-taped. Fieldnotes were taken during the discussion and written down after the discussion.

The conference series was approximately four years prior to the member-checking meeting, so participants were not asked for specific detail. A word-for-word transcript that provided a detailed reminder of the discourse was shared with each member checker. Specific questions regarding the results in chapter 4 were asked. The questions included the following:

- Does this comparative analysis between a "peak" session and "non-peak" session make sense to you and reflect the conference?
- Do you remember some sessions as being more engaging than others—or not?
- Did you feel that you could participate and share your thoughts during these sessions?

Specific questions about the results in chapter 5 included,

- Are there things that were said in this session that I am missing?
- What were your feelings when participating in this conference session?
- Do you identify with any of the participants mentioned?

- What are your thoughts about this analysis?

Each participant talked about how the description written in the chapter was accurate, but in different ways. Comments included "This was right," "this can be one fair way to represent [the session]," "this is accurate," and "you did a good job describing this." One person relayed that s/he was "transported" back to the conference series. Another person talked about having only "hazy" memories of this event because time had passed but that s/he "remember[ed the] tone" of the conversation.

Two member checkers talked about the importance of context and that the behind-the-scenes context of the conference series was not included in the findings. I shared that I had written a full chapter describing the history of the sponsoring organization, funding, numerous organizers, and context of conference series, and that this chapter was removed from the full document in order to ensure confidentiality of participants. For example, the important role that graduate students played in the organizing of this event is negated in this analysis. I concur: the member checkers raise an important point about context. To me as a researcher, balance between providing as much information as possible and not breaching confidentiality is a difficult and deeply noteworthy issue.

I also found that member checkers wanted to share their own narrative story about their individual roles in the conference series, what they saw as important in this conversation and how they felt about the interactions between people. There is power in the process of meaning making of an experience through narrative language and storytelling (Magolda & Abowitz, 1997; Nadeau, 1998; Smagorinsky, 2001). For example, one member checker remembered that it was never clear why this group was brought together for this conference series. S/he mentioned that there was discussion of building a movement, but nothing materialized from the discussions. Another person described the conflict, tension, and defensiveness between the organizer/s and the participants. Yet another person articulated the "us" versus "them" dichotomy between the community organizers and the academics in the room. I also learned that the direct comments made by "Glenn" (one of the invited speakers) during the large group discussion were intentional and planned prior to his attending the conference. Further, the member checker who helped with the conference materials walked me through all of the details leading up to the conference and described the behind-the-scenes conversations that were taking place during and after the conference series.

I also received conflicting advice from a few of the member checkers. One member checker talked about how it seemed that I was "pulling punches" with the analysis and that the discussion between participants during Session 5 seemed even more volatile than is depicted in the analysis. Another warned me about the political backlash that could possibly ensue from putting these findings in print, and that I should not attempt to critique too much at this early point in my academic career. My interpretation of each of these comments is that the member checkers sincerely cared about me and felt deeply about the potential influence this research could have (positive, negative, or otherwise), and that it was up to me to decide which course to take and whether or not I further revise the analysis and discussion. Further revisions of the analyses, with information from member checkers in mind, did take place.

The hours I spent individually with the member checkers was extremely valuable. Each had a different aspect of the story to share and each focused on a different part of the analysis—the aspect that was most salient to that individual. Information about who "poked" whom during the large group discussion to encourage verbal participation as well as the personal interpretations of the event helped add depth and credibility to the analysis. Interviews with participants, with the intention of using these interviews as data (not the intention here), would be beneficial in future studies.

Rich, Thick Description

Creswell (2003) states that researchers should "use rich, thick description to convey the findings. This may transport readers to the setting and give the discussion and element of shared experiences" (p. 196). In the analysis of the discourse between higher education leaders during the conference series on strengthening the relationships between higher education and society, I offer a rich and thick description of the discussion between participants. I also share much of the word-for-word transcript. If a comment by a participant is made immediately following another, I am clear to point this out so the reader may have a more complete understanding of the full discussion between participants. At times, participants comment on an idea shared much earlier in the session, or in a previous session. This is noted where relevant.

This rich, thick description may, at times, seem cumbersome, but all discourse represented in the analysis chapters are used in some manner. Further, I provide examples of each finding, as opposed to listing every performance move that represents a finding, so I may reduce the analysis to the most salient points while still providing a thick description of each example. In addition, I searched for disconfirming evidence with each theme in order to add to the trustworthiness of the findings as encouraged by Lather (2003).

Reflexivity

Research on matters of social identity and educational equity often engage scholars' own values and experiences. We all see, experience, and interpret the world through lenses and tools shaped by life in a racialized, gendered, and classed society. In this respect, a number of scholars argue that reflections and discussions of positionality can strengthen research, especially qualitative and interpretive research, and researcher reflexivity is often stressed as crucial to discussions of positionality (Jones, Torres, & Arminio, 2006; Lather, 2003; Salzman, 2002; Richardson & St. Pierre, 2005; Rossman & Rallis, 2003). For example, in an introduction to qualitative research, Rossman and Rallis (2003) talk about qualitative research as requiring four attributes, two of which emphasize self-reflexivity: (1) systematically reflects on who s/he is; (2) is sensitive to personal biography (p. 10). In addition, Milner (2007) argues that especially in the case of research involving issues of identity, "when researchers are not mindful of the enormous role of their own and others' racialized positionality and cultural ways of knowing, the results can be dangerous to communities and individuals of color" (p. 388) and to everyone involved. In a similar vein yet related to teaching, Friere (1970) notes, "liberationist teaching contains two dimensions: reflection and action, in such radical interaction that if one is sacrificed—even in part—the other immediately suffers" (p. 75). This connection between reflection and action is often found in emancipatory research in that the two are inextricably linked in order to make change. Richardson (2005) uses reflexivity as one of the criteria when reviewing papers or monographs submitted for social scientific publication as she wants to see a "social science art form" (p. 964) reflected in publications. In this way, reflexivity becomes an important publication process that in turn impacts tenure and promotion.

Each scholar provides a thoughtful and different rationale for why researcher reflexivity is important for quality qualitative scholarship. I have spent—and will continue to spend—dedicated time reflecting and writing on my own identities and roles as they relate to my approaches to this research study. I offer my role as a researcher with this particular study, rather than share all my reflexive musings on researcher reflexivity, as they may be found elsewhere. In those writings, I discuss how my own social identities (e.g., white, woman, Sicilian and Italian American) play a role in my approach to this study and in the analysis. Other fluid identities such as ability, age, class, sexual orientation, and spirituality also have an influence upon my lenses as a researcher.

I approach this study with the feeling that we will never understand the "whole" of the social construction of higher education's relationship with society, as the definition alters through time and is based on changing experiences and perceptions. As such, I do not offer a complete construction of the various frames and perspectives of higher education leaders, as the sociopolitical context and leaders' perceptions are constantly in flux—the current climate of drastic changes in the global economy and their direct impact on higher education is one example of factors causing such fluctuation. Instead, I present a snapshot of specific leaders' dialogic processes when discussing how to strengthen higher education's relationship with society in order to *begin* to consider the orienting questions. In addition, I approach research from the standpoint that objectivity is not a reality and offer this research not as "objective" truth but as one of many conceptualizations of this topic that will continue to change over time. With this in mind, I invite the coauthoring of articles, writing from alternative perspectives by participants of the events shared here, or publication in consecutive chapters of differing findings on the same aspect of study.

More concretely, I was not a participant or an observer to the first three conferences (of four) utilized in this research study. The focus of the pilot study was conference two and the focus for this larger study was conference three. I was a graduate student participant in some of the large group discussions in conference four. As a first-year graduate student research assistant, I worked with the authors of the final report of the conference series, not in terms of content, but in terms of working with the designer to move it through to publication and orchestrating the logistics regarding dissemination of the document.

It is important to note that the organizers did not fund this research study but willingly shared the conference audiotapes and other material. To me, this speaks to the organizers' willingness to open themselves up to scrutiny in order to contribute to the knowledge base and change around higher education for the public good—an important connection between language and action.

METHODS AND ORIENTING RESEARCH QUESTIONS

In order to triangulate the findings from the pilot study (Pasque & Rex, 2010) and literature review (chapter 2), and to further examine this topic in a "naturally occurring" context, I chose to explore discussions between higher education leaders during a national conference series using the emergent theoretical and methodological frameworks presented in this chapter. The initial orienting question proposes to uncover patterns that emerge from higher education leaders' discourse about how to strengthen the relationships between higher education and society: *How do higher education leaders talk about how to strengthen higher education's responsibilities to society? And, what are the implications of this discourse for higher education and society?* My goal in asking these questions is to use research to inform what we know in order to better transform it and to aid higher education leaders' ability to "inhabit the gap" (Komives, 2000) between theoretical language and daily action.

In this section, I describe the conference series, participants, and conference material. For the purposes of confidentiality, much of the context and background narrative of this conference series is intentionally absent. In addition, I describe the specific content of comparative analysis (Strauss & Corbin, 1999) methods adopted for identifying "peak" conference sessions. The "peak" sessions, or sessions where there was more conversation between participants (as opposed to sessions that were limited primarily to participants asking questions and the main speaker answering them), were identified in two different ways. Each method of conversation analysis (language as sequence and structure) used in chapter 4, the first of the analysis chapters, is described. Finally, I share the critical discourse analysis and multidimensionality processes (language as constructed and shaped by social constructs) used for chapter 5, the second analysis chapter.

The National Conference Series

This study explores a national conference series held during the early twenty-first century in the United States. A national higher education organization, in conjunction with a foundation, sent out invitations that participants could accept or decline. The organization funded travel, food, and lodging for participants. Leaders included representatives from national foundations, national associations, and state legislators as well as university presidents, faculty, student affairs administrators, graduate students, community partners, and a few undergraduate students. The series was "naturally occurring material"

(Peräkylä, 2005, p. 869), that is, it was not specifically designed for a research study. Such naturally occurring interaction addresses Sacks (1964) critique of the use of interviews:

> The trouble with [interview studies] is that they're using informants, that is, they're asking questions of their subjects. That means that they're studying the categories that Members use...they are not investigating their categories by attempting to find them in the activities in which they're employed. (as cited in Roulston, 2004, p. 151)

I argue that although interviewing plays an important role in research, one of the strengths of this study is that it focuses on naturally occurring interaction among and between higher education leaders in order to explore the meaning made by the participants in a real-life setting within the boundaries of the particular conference series.

This organization and cosponsoring organizations took the initiative to gather higher education leaders for this national conference series in order to try to strengthen the relationships between higher education and society. Although the intent and motivation of the organizers are not the focus of this study and should be further explored through future research studies, it is important to note that this organization and many other organizations across the country are working on this difficult topic and trying to strengthen the relationships between higher education and society on a regular basis. Members of this specific organization initiated the idea, implemented this national conference series, and, since, spearheaded a number of subsequent research studies, policy briefs, gatherings, and programmatic efforts in order to try to further these efforts. As mentioned, it is through the organization's commitment to furthering knowledge and information on this topic that they agreed to participate in this research study and contribute toward making change.

Curriculum Design Team and Participants

Thirteen participants served on the curriculum design team. The curriculum design team included three African American people, ten white people, seven women, and six men. The team included people who held fairly prominent positions in various types of higher education associations, organizations, universities, and foundations. There were two emeritus professors, one full professor, one associate professor, five presidents or co/directors of organizations, three vice presidents or associate directors of universities or national associations, and one foundation program officer. The areas of interest

Table 3.1 Matrix for Selection of Participants

	Senior	*Mid-Career*	*Early-Career*
Community Leaders			
Faculty			
Foundation Program Officers			
National Associations			
State Leaders			
Students			
University Presidents			

of the curriculum design team participants were many—citizenship, diversity, democracy, academic affairs, and leadership, to name a few. The curriculum design team also represented different regions throughout the United States.

Site organizers and facilitators were hired to coordinate each of the four different conferences in the series. The curriculum design team intentionally considered whom to invite to the various conferences. The goal was to equally represent late career (senior or emeritus scholars and administrators), mid-career, and early career (participants likely to be engaged fully in this work in the mid-twenty-first century)—a third of the participants for each career phase. More specifically, the design team created a participant selection matrix that consisted of various characteristics that would assist in the identification of invitees from different categories. See table 3.1. As invitees RSVP'ed "no," another invitation was issued in its place, usually to a person who was located on the same place in the matrix. In addition, the team tried to pay attention to diversity across gender and race. This strategy provided a pool of participants with the diverse characteristics deemed important by the curriculum design team.

The series of four, three-day conferences together included over 200 participants. Most participants and the curriculum design team attended two conferences in the series, including the final conference. Many of the faculty and staff at the organizing office attended all four conferences.

The race, gender, class, and position of participants were gathered through self-reporting mechanisms during the conference itself (e.g., "I am a black man"), through informal discussion with organizers, or, as in the case of position descriptions, through self-report to the organizers of the series. In two cases, the member checkers provided additional information about their social identity. When the social identity of a participant is not self-reported, I do not make assumptions about the participant.

Conference Material

Prior to each conference, participants received a letter of invitation and, once accepted, they were mailed a three-ring binder with context material. This context material included research articles, a list of invited participants, and travel information. The topics of the research articles include civic engagement, democracy, higher education for the pubic good, and other related issues. In addition, a postdoctoral researcher and a graduate student gathered information to create annotated bibliographies on the topics of civic engagement, equity, and service-learning; copies of this compilation were provided at the conferences.

At each conference, there were large posters placed around the room that displayed different figures, each a theoretical basis for the discussion. One figure was the Ecological Impact Model, a standard ecological model that circularly represented four different levels: the individual, university, system, and the public. During each introduction to the conference, the organizers pointed out the figure and highlighted the crescent between the university and system levels as the focus for the conference.

The second figure around the room was the Dialogic Process Model. This model showed the process of dialogue as a circular movement from "awareness" to "understanding" to "commitment" to "action." The organizers also discussed the model as a principle of the convening, with the goal of the conference series as a movement from awareness to action. One element of action was a culminating final report and a common agenda.

Prior to participation in the conference series, participants were asked to define the phrase "higher education for the public good" and return that definition to the organizers. A compilation of all definitions was provided in the packet of materials for the conference. In the opening session of the first conference, one of the organizers reflected on the variety of these definitions. A full discourse analysis on these definitions may yield additional information. The packet also included an agenda, a list of participants, and related research articles chosen by the facilitators and organizers.

Data Reduction

Four, three-day conferences were considered for this study. See table 3.2 for an outline of all the large group conference sessions. A

Table 3.2 Outline of Formal Agendas for the National Conference Series

	Description of Sessions			
	Conference I	*Conference II*	*Conference III*	*Conference IV*
Day 1	Reception and Comments Welcome and Introduction	Reception and Welcome Keynote Speech	Reception and Welcome Comments (Keynote)	Opening Session Reception & Remarks
	Keynote Speech and Discussion	Conceptual Framework	Setting the Context	Welcome & Reflection
	Conceptual Framework	Introduction	Creating Culture Change in the System of Higher Education	Setting the Context
Day 2	Public Understanding and Support: The Covenant between Higher Education and Society	What Does It Mean to Educate for Social Responsibility?	Creating Culture Change in Colleges and Universities	Concurrent Sessions (Participants could choose from six)
	Breakout Session	Breakout Session	Breakout Session	Concurrent Sessions (Participants could choose from seven)
	Presentation	The Problem of Culture	Reflections	Process for Crafting the Common Agenda
	The Role of Public Policy and the Covenant between Higher Education and Society	Breakout Session	Creating Significant Change in the Campus Culture	Breakouts by Affinity (job/ position related groups)

	Breakout Session	What Constitutes Good Practice in Educating for the Public Good?	Breakout Session	Building our Collective Commitment to Next Steps (included paper presentation of affinity group points & looking to the future)
Day 3	Reflection of Breakouts from Graduate Students and Discussion	Breakout Reports	Creating Change in the Disciplines	Closing Presentation (Lecture open to the public)
	The Road Ahead	How Can Our Deliberations be Translated into Policies and Actions?	Reflections	
		What Issues Should Be Given Priority at the Fourth Dialogue?	Building Alliances for System-Wide Change	
			Breakout Reports The Road Ahead Closing Remarks	

total of 48 formal large and small group sessions lasted between one and two hours a piece. Each session was audio-recorded. A professional journalist attended all of the conferences, created a scripting of each dialogue, and authored final reports,[2] copies of which each participant received. Conference materials, audiotapes, and scripts were utilized in the constant comparative method of this study (Charmaz, 2005; Miller & Fox, 2004; Strauss & Corbin, 1999). I read through all conference material and listened to all large group sessions during the conference series. I wrote memos and notes on copies of the conference agendas in order to keep track of the main crux of each session. In addition, I created a context narrative of each conference session, so I would not lose the focus of the larger conference discussion as I drilled down to more specific dialogic exchanges.

As mentioned, a site organizer and a facilitator were hired for each of the four conferences. One of the facilitator's roles was to keep track of who wanted to speak and maintain the order of speakers during group discussions. In a number of sessions, one or two invited speakers would address the group and then a large discussion would be facilitated by the identified facilitator. Many of the comments in the large group discussion would be addressed directly to the invited speaker, whereas at other times, the group would break away from the speaker/participant question-and-answer type session and move to a more of a dialogic format of interaction between participants. This dialogic interaction was the focus of this study and is identified through specific methods described below. However, even during these more dialogic moments, the facilitator continued to keep track of the order of speakers, so—at times—the discussion is not quite linear and people "come back to" an earlier point. In the rich, description of the findings, I make certain that the specific order of the discussion is highlighted for clarity and analysis purposes.

IDENTIFYING THE "PEAK" CONFERENCE SESSIONS

The "peak" sessions, or sessions where there was more conversation among participants—as opposed to sessions that were limited primarily to participants asking questions and the main speaker providing answers—were identified through two different processes, each of which is described in this section. These two processes yielded similar results and add to the trustworthiness of this study.

Two different techniques were applied to pinpoint the telling narrative interactions for closer analysis of the "peak" sessions (Mitchell, 1984). First, I listened to the audiotapes from the conferences and identified key concepts mentioned most frequently. For example, when considering the second conference, "Higher education for the public good" (HEPG) was mentioned most often, as might be expected (Pasque & Rex, 2010) (See table 3.3). Second, the number of times per session that participants specifically mentioned the phrase "HEPG," or a form thereof, was counted and graphed (See figure 3.2). Figure 3.2 illustrates why session five, in which HEPG was mentioned far more often than in other sessions, was selected as the most promising material. The findings and analysis were discussed regularly with a research team over the course of four months. We (Pasque & Rex, 2010) attribute the dramatic increase in frequency to the preceding sessions that built toward this climactic session.

The second process in which "peak" sessions were identified started with my listening to all of the large group sessions and reflecting on identifying the sessions where discussion occurred between participants. I identified "peak" sessions as containing more probing and engaging discussion between participants; there was more depth to the conversations between participants than in the guest speaker-participant questions-and-answer sessions.

My intuitive findings for the first three conference sessions matched the findings for the second conference: The "peak" sessions were consistently on the third day during the session immediately

Table 3.3 Frequency Count of Words/Concepts Discussed during Conference II

Word	Frequency
Higher Education for the Public Good	59
Change (change models, systemic change, change agents, changing lives)	38
Dialogues	20
Power (power, powerless, power differentials)	19
Leadership	18
Movements	17
Democracy	17

Note: The frequency is not the actual number of times a word was mentioned but the number of times the topic was talked about or brought up in conversation by an individual speaker. The actual word may have been utilized 1 to 3 times during the speaker narrative, but the occurrence was coded only once. Thus it prevented increased weight from being given to speakers who used the word repetitively versus others who referred to "it."

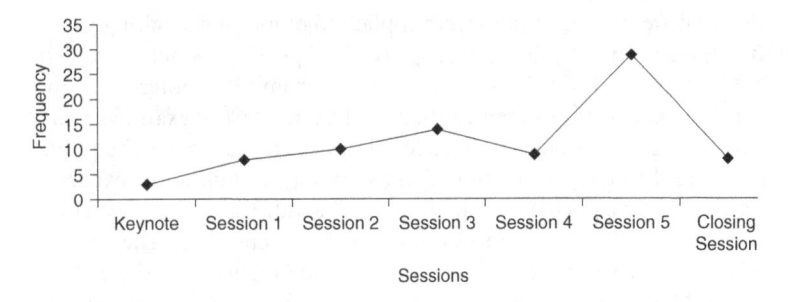

Figure 3.2   Frequency Count of "Higher Education for the Public Good" during Conference II

following the "report out" from the small group discussions. The fourth conference did not have a "peak" session as the organizational structure was different and did not allow for a large group discussion between approximately 200 participants.[3]

Each of the "peak" sessions was transcribed word-for-word. I identified participant names from the facilitator comments. At times, a participant was unknown. In order to identify the participants that were unknown to me, a participant in the conferences and research assistant from the sponsoring organization reviewed the transcripts and listened to the audiotapes. She identified additional participants. At this point, I shared the transcripts and audiotapes with one of the main organizers, and he was able to identify the remainder of the participants whose identity was unclear. Neither of these people served as a member checker in order to keep the roles separated.

In order to increase credibility (Lincoln & Guba, 1985), I compared the similarities and differences between "peak" and "non-peak" sessions through a constant comparative method (Strauss & Corbin, 1999) that was fairly detailed. In this way, I support my intuitive findings with a more post-positivist process (findings in chapter 4) prior to moving on to the next stage of the research process (findings in chapters 5 and 6). For this process, I chose to focus on the third conference as the second conference was already explored in the pilot study and the first conference (of any conference in a series) was perhaps more likely to have logistical problems with the potential to interfere with conference participation.

The outline of the formal agenda for Conference III is listed in table 3.4. Session 5 was intuitively identified as the "peak" session. With a roll of the dice, I randomly chose to examine Session 3 as the

Table 3.4 Outline of Formal Agenda for Conference III

Day	Session Type	Description of Sessions
1	Introductory	Reception and Welcome
	Introductory	Comments (Keynote)
	Introductory	Setting the Context
2	Plenary and Discussion 1	Creating Significant Cultural Change in the System of Higher Education
	Plenary and Discussion 2	Creating Significant Cultural Change in Colleges and Universities: The Lessons from an Accumulation of Practice-Based Wisdom
	Small Groups	Breakout Group Session
	Reflective	Reflections on the Morning Work
	Plenary and Discussion 3	Creating Significant Change in the Campus Culture
	Small Groups	Breakout Group Session
	Plenary and Discussion 4	Creating Significant Cultural Change in the Disciplines
3	Reflective	Reflections on the Yesterday's Work [*sic*]
	Plenary and Discussion 5	**Building Strategic Alliances for System-Wide Change: An Exchange**
	Reporting	Breakout and Plenary Reports
	Reflective and Closing	Reflections and the Road Ahead
	Closing	Closing Remarks and Adjourn

comparative "non-peak" session. I then proceeded to compare the two sessions by examining each participant's performance moves utilizing numerous discourse theories as tools for analysis. The structural and functional performance moves for Session 3 and 5 are in tables 3.5 and 3.6.

The tools for analyzing performance moves (as found in the discourse analysis and communication literature) are described in detail in the next chapter. After coding the tools for analysis, additional categories were added and coded for including "participant asks a question of the organizer/facilitator," "facilitator asks a question," "organizer asks a question," and the like. This approach to identifying the "peak" session supported the intuitive findings that the large group session on the third day, immediately following the small group report out, was the most dynamic and engaging session and, therefore, was selected for further analysis. The results of the coding for the tools of analysis throughout Sessions 3 and 5 are the focus of the next chapter.

Table 3.5 Conference III, Plenary Session 3: Turn Taking and Performance Moves

Turn No.	Person Number (Structural)	Moves (Functional)
1.	Person 1/Invited Speaker	Invited Plenary Speech—African American man from a university
2.	Person 2/Invited Speaker	Invited Plenary Speech—Woman student from a technical college
3.	Person 3/Organizer	Asks a question of person/speaker 1
4.	Person 1/Invited Speaker	Answers question with a story*
5.	Person 4	Asks a question of person/speaker 2
6.	Person 2/Invited Speaker	Answers question with a story*
7.	Person 5	Links comments from speakers 1 and 2 with a speaker earlier in the day (person 9)
8.	Person 1/Invited Speaker	Answers with a story* about the structural diversity of faculty
9.	Person 6/Facilitator** of this Session	Asks a question of person/speaker 2
10.	Person 2/Invited Speaker	Answers with a story*
11.	Person 7	Aligns with both invited speakers (persons 1 and 2)
12.	Person 8	Aligns with person/speaker 1; shares story; asks question about her own story (not directed to speakers or participants)*
13.	Person 7	Shares story*
14.	Person 9/Invited speaker earlier in the day	Shares story*
15.	Person 2/Invited speaker	Shares story*
16.	Person 10	Asks question of person/speaker 1
17.	Person 1/Invited Speaker	Answers question with a logical syllogism ($x + y = z$; $a = z$; $x + y = z$)
18.	Person 11	Comments to person/speaker 1 about the logical syllogism (not a question, alignment, or disagreement)
19.	Person 1/Invited Speaker	Aligns with person 11
20.	Person 12	Asks question of person/speaker 2
21.	Person 2/Invited Speaker	Shares story*
22.	Person 6/Facilitator	Asks a follow-up question to person/speaker 2

Continued

Table 3.5 Continued

Turn No.	Person Number (Structural)	Moves (Functional)
23.	Person 2/Invited Speaker	Responds to the question
24.	Person 13	Asks question to both the speakers (persons 1 and 2)
25.	Person 6/Facilitator	Picks person 14 to answer this question
26.	Person 14	Answers question (from person 13) with a story
27.	Person 13	Asks a follow-up question
28.	Person 14	Responds to the follow-up question
29.	Person 15	Shares story*
30.	Person 16	Asks a question of person/speaker 1
31.	Person 1/Invited Speaker	Responds with a story*
32.	Person 16	Asks a clarifying question
33.	Person 1/Invited Speaker	Shares a response story*
34.	Person 17	Shares comment with person/speaker 1
35.	Person 1/Invited Speaker	Response to comment
36.	Person 17	Asks person/speaker 1 "yes" or "no" question
37.	Person 1/Invited Speaker	Responds "yes"
38.	Person 3/Organizer	Asks a question to person/speaker 1
39.	Person 1/Invited Speaker	Aligns with person 3; answers with a story*; person 18 is mentioned in story
40.	Person 18	Shares story* that includes a logical syllogism ($x + y = z; a = z; x + y = z$)
41.	Person 19	Shares story* that relates to person 18's story
42.	Person 3/Organizer	States that "something is missing" from these large group conversations (in role of organizer)

* A story is a personal narrative of one's own experience at an institution or with a particular program.
** Facilitator is mentioned when s/he added more to the conversation rather than just calling on the next speaker.

The peak session entitled "Building Strategic Alliances for System-wide Change: An Exchange" was selected for a more in-depth critical qualitative analysis. There were 56 participants who attended this session, 30 of whom spoke at least once during a total of 86 performance moves. See table 3.6 for a complete list of performance moves made by each participant during this session.

Table 3.6 Conference III, Plenary Session 5: Turn Taking and Performance Moves

Turn No.	Person Number (Structural)	Moves*** (Functional)
1.	Person 1/Invited Speaker	Invited Plenary Speech—white woman, director of national organization
2.	Person 2/Invited Speaker	Invited Plenary Speech—African American man, community partner
3.	Person 3	Aligns with both speakers and added a new idea
4.	Person 4	Disagrees with speaker; questions organizers' model
5.	Person 2/Invited Speaker	Disagrees with person 4
6.	Person 5	Shares a story* that relates with person/ speaker 2
7.	Person 6	Aligns with speaker and person 5
8.	Person 2/Invited Speaker	Asks a question about the meaning of something person 6 said
9.	Person 6	Apologizes and describes what it is
10.	Person 7	Aligns with person/speaker 2 and shares a new idea; asks a question of person/ speaker 2
11.	Person 8	Shares a story*; humor
12.	Person 9	Shares a story*
13.	Person 10	Asks a question of another participant
14.	Person 11	Shares a story*
15.	Person 12	Aligns with person 3; humor
16.	Person 13	Shares a story*
17.	Person 14/Facilitator** of this session	Asks a question of person 13
18.	Person 13	Responds to question (without alignment/ disagreement or story)
19.	Person 15	Aligns with person/speaker 1; aligns with participant 12
20.	Person 16/Organizer	Situational power; shapes the agenda
21.	Person 17	States new idea; asks question of the group of participants
22.	Person 18	Aligns with person/speaker 1; asks a question of the group of participants
23.	Person 19	Footing, where positions self "up" (as better than others); shares a story*
24.	Person 3	Disagrees with person 19
25.	Person 20	Corrects person 3 (situational power)
26.	Person 3	Thanks person 20; finishes idea...
27.	Person 14/ Facilitator**	Interrupts person 3; calls on person 21 and states that s/he is a "student" (situational power)

Continued

Table 3.6 Continued

Turn No.	Person Number (Structural)	Moves*** (Functional)
28.	Person 21	States that s/he is not a student
29.	Person 14/Facilitator**	Person 14 may have apologized here, but inconclusive as a few people are talking at this moment
30.	Person 21	Aligns with both speakers/persons 1 and 2; uses verbal art; shares a new idea; humor; ends by footing where positions himself "down"
31.	Person 14/ Facilitator**	Talks about power; "negative face"
32.	Person 22	Aligns with person 21; shares a story*; aligns with person/speaker 2; asks question of organizers (gives organizers power)
33.	Person 1/Invited Speaker	Aligns with participant 21
34.	Person 23	Shares a story*
35.	Person 14/ Facilitator** of dialogue	Echoes a point that person 23 made
36.	Person 23	Responds
37.	Person 24	Aligns with person 21; disagrees with the organizing (systems) model; mentions other models
38.	Person 14/ Facilitator**	Calls on person 4
39	Person 4	States that he wants to speak to a different point than what person 24 brought up; asks the group if someone else wants to speak to the point that person 24 made
40.	Person 14/ Facilitator**	Asks what topic was on the floor
41.	Person 16	Responds to person 24; frames person 24's comment as a question; states that he hopes her question is rhetorical and will not be answering this question at the moment (situational power)
42.	Person 24	States to person 16 that answering such a question would be another conference
43.	Person 16	Agrees with person 24 that it would be another seminar
44.	Person 24	Tells person 16 that one example would be helpful and stops talking as person 16 starts to talk
45.	Person 16	Interrupts and talks about "values" and "working" not "models," which was person 24's original statement; persuasion—quasilogical

Continued

Table 3.6 Continued

Turn No.	Person Number (Structural)	Moves*** (Functional)
46.	Person 4	Talks about social movements; asks a question of the group about one of the elements in the model; negative face
47.	Person 25	New idea; challenges status quo of group conversation; persuasion as a verbal art; footing aligns down; other participants give verbal support; shares story*
48.	Person 14/ Facilitator** of dialogue	Encourages a person to speak (who does not end up speaking in this dialogue); calls on person/speaker 2
49.	Person 2/Invited Speaker	Disagrees with participants (there is no social change model); talks about institutional power; other participants give verbal support
50.	Person 26	Interrupts to ask a question of person/speaker 2
51.	Person 2/Invited Speaker	Asks a question back to person 26
52.	Person 26	Makes a short statement in response
53.	Person 2/Invited Speaker	Asks a question of person 26; Footing aligns self up
54.	Person 12	Disagrees with person/speaker 2
55.	Person 2/Invited Speaker	States disagreement with participant in a different way
56.	Person 12	Disagrees with person/speaker 2 again: "that's not true"
57.	Person 2/Invited Speaker	Tells person 26 to identify an example of hir point
58.	Person 12	States that s/he will later and is not interrupting now (deference)
59.	Person 2/Invited Speaker	States point again
60.	Person 16	Disagrees with person/speaker 2; formal politeness
61.	Person 2/Invited Speaker, Person 14/Facilitator** and Person 16/ Organizer	All talking at once
62.	Person 14/Facilitator** of dialogue	Prevails; states to person 16 that there is a second point that person/speaker 2 wants to make
63.	Person 2/Invited Speaker	Makes comment
64.	Person 16/Organizer	Aligns with person/speaker 2's point from the day before; states that the point from the day before is what s/he (person 16) was trying to do through his words

Continued

Table 3.6 Continued

Turn No.	Person Number (Structural)	Moves*** (Functional)
65.	Person 2/Invited Speaker	Statement about process (code switching); Statement about institutional power; persuasion presentational—verbal art; makes comment about own social identity (race)
66.	Person 21	Challenges something person/speaker 2 said
67.	Person 2/Invited Speaker	Rejects what person 21 stated
68.	Person 21	Disagrees with person/speaker 2; aligns with person/speaker 2; solidarity with person/speaker 2 (race)
69.	Person 2/Invited Speaker	Face threatening Act—save face; solidarity with group (academics); tries to increase own positive face
70.	Person 27	Face threatening Act (whispering "he's lost it")
71.	Person 14/Facilitator**	Comment to person/speaker 2; adds idea
72.	Person 2/Invited Speaker	Agrees with person 14/facilitator** comment
73.	Person 28	Face threatening Act
74.	Person 29	Asks for a repeat of person 14/ facilitator's** comment
75.	Person 14/Facilitator**	States it again
76.	Person 12	Aligns with person/speaker 1; disagrees with person/speaker 2; politeness move of camaraderie; shares story*; humor; footing
77.	Person 2/Invited Speaker	Politeness move—formal
78.	Person 12	One-word comment
79.	Person 2/Invited Speaker	Adds another question to the topic at hand
80.	Person 12	States that s/he knows what person/speaker 2 stated
81.	Person 2/Invited Speaker	States hir feeling
82.	Person 12	States hir feeling
83.	Person 30/Organizer	Aligns with person/speaker 2 and person 21; shares story*; aligns with person/speaker 2; face threatening Act—gives person a way to "save face"
84.	Person 2/Invited Speaker	States his point again (does not take the "out")
85.	Person 30/Organizer	Describes point again
86.	Person 2/Invited Speaker	States that he understands person 30's point; code switching; situational power

* A story is a personal narrative of one's own experience at an institution or with a particular program.

** Facilitator is mentioned when s/he added more to the conversation rather than just calling on the next speaker.

*** Moves provided in this table are primary or the major moves made by the participant comments but do not include all coded tools for analysis. For all tools for analysis for Session 5 see table 4.2.

An In-depth Analysis of the "Peak" Session

My focus is on the process and content of the discussion between higher education leaders. In K-12 literature, various perspectives for considering classroom interactions have been identified (Rex, Steadman, & Graciano, 2006). For the purposes of this study, I take what Rex, Steadman, and Graciano describe as a "sociolinguistics and discourse analysis perspective" where I forward the concept of reflexivity of language. More specifically, as participants interactively construct the social fabric of the dialogue through discourse, the discourse shapes participation in the discussion as well. Stated another way, *the discourse is shaped by context as it shapes its context.*

At this point, a thematic analysis utilizing a constructivist grounded theory approach (Charmaz, 2005) was conducted using the transcript for Session 5, the session identified as a "peak" session. I read through the transcript a number of times and emically coded the themes that emerged from the transcript. A number of emergent themes were shared with research colleagues at my university and at a national conference presentation, and feedback was provided. "Narratives that challenge or resist the dominant perspective" emerged as one of the primary themes worth exploring further, and thus the scope of this research study became more defined.

From here, I explored the theme "narratives that challenge or resist the dominant perspective" across horizontal and vertical dimensionality (Swanson, 2006; also see Bautista's 2006 research on punks, darks, and rockabillies), which is also referred to as intertextuality (Johnstone, 2002). Horizontal repertoires are performance moves that alter the trajectory of the discussion. Horizontal repertoires move across topic and add to the breadth of the conversation. As each topic (horizontal repertoire) is discussed in more detail, participants add to the depth of the conversation, or "vertical repertoire," with their performance moves. Throughout this analysis, participants move across horizontal and vertical repertoires, adding to the multidimensionality and complexity of the socially constructed dialogues.

There were two horizontal moves during Session 5 that challenge or resist the dominant perspective: challenges or resistance to (1) the models and programs presented by conference organizers in earlier sessions, in handout material, or on large posters placed around the room or (2) the content of the dialogue engaged in by participants and organizers. This observation is not to say that the dominant models were or were not crafted by feminists, post-Marxists, or people who did not care about strengthening the relationships between higher education and society. It is to say that participants in two horizontal moves challenged the models that were presented at the conference (and this finding supports earlier and independent findings from the pilot study). Importantly, each of these challenges to the models was then followed by a number of different comments where participants aligned/disagreed with the originator of the comment or aligned/disagreed with the dominant ideologies. The follow-up comments add to the vertical dimensions of the conversation—often deepening the conversation with every verbal comment. For example, when a participant aligns with a participant who spoke earlier by saying, "I agree with what Glenn said and...," their comment adds to the vertical dimensions of the repertoire initially offered by Glenn.

In the fifth and sixth chapters, I focus on the multidimensionality of the theme "narratives that challenge or resist the dominant perspective." This approach to exploring the content of the dialogic process adds to the depth of the analysis as well as provides an additional tool for analysis to understand the complexity of the conversation. I also iteratively consider this discussion through the theoretical framework where I explore the ways in which participants' social identities, or the social identities of people who are the topic of discussion, play a role in the conference series.

After the emergent themes evolved and an intertextuality analysis was conducted, I reviewed the final reports of the conference series and conducted a line-by-line content analysis (Marshall & Rossman, 1999). The information gathered from this content analysis was compared and contrasted to the findings from the discourse analysis of the conversation during the sessions.

In the next chapter, I share the details of the conversation analysis that led me to identify the "peak" sessions. In the fifth chapter, I focus in on the "peak" sessions and offer a detailed critical discourse analysis of both the process and content of the discussions as I consider these findings with the theoretical lenses identified. In the sixth chapter, I provide further depth to this complex analysis by narrowing the theoretical focus to that of a feminist perspective. Finally, I reflect on the entire set of findings from the pilot study, literature review, and larger study in order to offer new knowledge that may be transformative to higher education policy dialogue and action. As each iterative chapter provides further depth to this critical analysis in the hopes of making concrete change, the words of Norman Denzin and Michael Giardina (2009) from *Qualitative Inquiry and Social Justice* are called to mind:

> This is a historical present that cries out for emancipatory visions, for visions that inspire transformative inquiries, and for inquiries that can provide the moral authority to move people to struggle and resist oppression. The pursuit of social justice within a transformative paradigm challenges prevailing forms of human oppression and injustice. (p. 11–12)

CHAPTER 4

Behind Closed Doors: Communication Theories and Conversation Analysis

Have you ever participated in a conference and continued to mull over the discussion, even days after you returned home? Have you witnessed or experienced interactions between people that were difficult to describe and left you with an uncomfortable pit in your stomach? What is different between these types of conferences and those that leave you with a feeling of having accomplished something or engaged in meaningful dialogue? Do these same complex struggles happen between leaders in national higher education policy discussions?

It is not often that we have such an opportunity to go behind closed doors of national policy conversations to hear what happens when leaders discuss difficult issues. This micro-analysis chapter begins to break down some of the interactions between higher education leaders as they engage in dialogues at a national conference series. This process is important for higher education leaders as it provides us with tools to dissect intergroup dynamics in "naturally" occurring contexts (Peräkylä, 2005, p. 869). Notably, if the complexities of intergroup dynamics go unacknowledged, then they will continue unaddressed. Left alone, negative dynamics or unclear messages may hinder our work toward strengthening the relationships between higher education and society and efforts toward educational equity and social justice.

In order to understand the complexities of how higher education leaders talk about concepts of higher education for the public good and how leaders' cognitive models operate in the conference series, it is necessary to select a session with an interaction that affords a rich example. In this chapter I explain how I theoretically and strategically selected a session of dynamic group interaction involving multiple participants for further in-depth analysis in subsequent chapters. From this point forward, I will call this a "peak" session (which contained more dialogue between participants) and compare it to a "non-peak" session (sessions that were limited primarily to participants asking questions and the main speaker giving answers). I also explain the various communication theories guiding my conversation analysis and how they apply to the "peak" and "non-peak" sessions. I lay out a procedural trail for the reader to follow and, in doing so, indicate how my choices strengthen the trustworthiness of this study. It is important to note that conversation analysis is "a markedly 'data centered' form of discourse analysis: in 'pure' versions…the analyst is not supposed to appeal to any evidence that comes from outside the talk itself" (Cameron, 2001, p. 87). Therefore, I use conversation analysis in this chapter to identify "peak" sessions, which I will explore more deeply in the next chapter using critical discourse analysis that takes into account how reality is constructed and shaped by various social forces (Cameron, 2001).

I begin by focusing on Conference III's large group sessions, which include the introduction, context session, plenary and discussion sessions, small group sessions, reflective sessions, and a closing session for a total of fifteen sessions. After listening to audiotapes of all large group sessions, it became evident that the discussion among participants was most engaging in plenary and discussion Session 5. In this chapter, I explore the similarities and differences between plenary Session 5 and other plenary sessions in order to show why the former was the most dynamic session. Specifically, I share a brief definition of each communication theory used as a tool for analysis and offer an example of each, as found in the field of higher education. Finally, I compare and analyze "peak" and "non-peak" plenary sessions by using the various tools for analysis commonly found in conversation analysis and interpersonal communication. These micro-analyses help determine that Session 5 is the best session to consider when talking about strengthening the relationships between higher education and society, which I explore in depth in the next two chapters.

DESCRIPTION OF TOOLS FOR ANALYSIS FROM COMMUNICATION THEORIES

This section consists of a brief description of the tools for analysis used in communication theory and conversation analysis (see table 4.1). I give examples of praxis in higher education to further describe each concept. Each tool for analysis offered in this section will assist in the exploration of units of analysis—or the "level of inquiry on which the study will

Table 4.1   Tools for Analysis

Agenda's—Creation of
- Audience Design Theory
(design language to address audience in the room and not in the room)
- Code-Switching
(back and forth between different styles
of language across race / ethnicity / class)

Face Threatening Acts (FTA)
- Positive Face
- Negative Face

Hedging (um, um, I, I)

Higher Education for the Public Good:
- as Object (bodies, illness, death)
- as Subject (doctor, patients, nurses)

Interactions
- comment on what the speaker said (agree / align)
- comment on what the speaker said (disagree / reject)
- comment to others in the room
- question to the speaker

Performances of Identity (identity as presentation of the self)
- Race & Ethnicity
- Gender

Persuasion
- Logical syllogism (x + y = z; a = z; x + y = a)
- Quasilogical (person a likes b; b likes c; so a will like c; relations are not always
logical)
- Presentational (rhythmic flow of words & sounds—poetry), also—Verbal Art
- Use of stories

Politeness—Rules of Politeness
- Formal (distance)
- Hesitancy (deference)
- Equality (camaraderie)

Power & Control

Power & Solidarity
Power
- Institutional Power
- Situational Power
Solidarity
- Solidarity
Footing—Participant states something that causes a shift in position/power either
up or down

Question to the Speaker

Silence

Structural conventions

Continued

Table 4.1 Continued

Turns (Structural)

Moves (Functional)

Narrative Chunks (so, because, therefore, and)

Verbal supportive gestures "um hmm" or "uh, huh"

Whose words / topics get used, picked up, and whose do not

Total number of comments

Total number of comments by different people (some people spoke more than once)

---

focus" (Marshall & Rossman, 1999, p. 34)—namely individuals' verbal narratives shared during a large group conference session.

Code-switching or *style-shifting* happens when a person shifts back and forth between two different styles of language or ways of speaking. For example, Gumperz's (1982) research addresses the verbal and nonverbal code-switching of an African American politician who shifted between formal and informal vernacular during a speech he made in an African American protestant church and explores why this code-switching was not well received by a particular congregation. In a college or university setting, code-switching may occur, for example, when an undergraduate uses with hir friends a style of language that is unfamiliar in their home environment. The undergraduate may choose to use a different verbal style with faculty members in class than he/she uses with friends from high school. This ability to code-switch helps a person to tailor a message to a particular audience and adapt to various environments and contexts.

A *face threatening act* (FTA) is verbal language that poses a threat to a person's face—positive (a person's desire to be approved of) or negative (a person's desire to have unimpeded actions) (Brown & Levinson, 1987). For example, in a college setting, if a trustee is speaking in a negative manner about the college president, then this verbal language poses a threat to the college president's positive face. If a trustee is critiquing the college president's decisions and actions regarding the hiring process for the vice president for research, it poses a threat to the college president's negative face. The word "face" is similar to the familiar expression "saving face"—working to reduce or eliminate the threat to one's positive face.

Hedging happens when a person repeats a word more than once before stating the remainder of the sentence (Johnstone, 2002). Hedging is distinct from stuttering. When a person says "I, well I, I think that..." before making a statement, he is hedging. There are a number of different reasons for hedging. Hedging could provide a verbal stall in the conversation and help provide the person with enough time to think through a response. Hedging could indicate doubt or accompany a response that is in disagreement with what another person believes. Hedging may also accompany shifts in power. For example, an administrator verbally stating her disagreement with her department chair may use hedging as a way to soften the impact of the opposition.

Humor may be used for a number of different reasons such as to minimize (DeVito, 1992; Tannen, 1994), to address an FTA, or to build camaraderie. Gronbeck, McKerrow, Ehninnger, and Monroe (1990) suggest that one uses humor to make people feel connected, deflect hostility, or offer unpalatable or biting critique. For example, when the dean of the music school delivers a speech to potential donors, she may use humor to connect with the audience and make them feel connected to her, the institution, and the need for renovations to update the building with current technology, performance space, and soundproofed practice rooms.

Interactions among participants identify from whom and to whom a comment is addressed (Edwards & Potter, 1992). For the purposes of this study, a number of different types of interactions were considered. These include

1. interactions with the facilitator (aligns with participants, asks a question),
2. interactions with participants (dis/agree with speakers, align then disagree with speakers, repeat what speaker stated and add a new idea, ask a question of the speaker, dis/agree with a speaker from an earlier point in the discussion, dis/agree with other participants in the room, ask a question of the facilitator, organizer, or another participant),
3. interactions with the speaker (dis/agree with participants, repeat what a participant stated and add a new idea, ask a question of the participant or facilitator), and
4. interactions with the organizer (dis/agree with speaker, ask a question of the speaker, shape the agenda or direction of the discussion).

Social identity is defined as a person's ability, age, class, ethnicity, gender, gender expression,[1] race, religion, and sexual orientation (Hardiman & Jackson, 1997). Targets are members of social identity groups that are disenfranchised, exploited, and victimized through the social, cultural, and historical reproduction of power. Targets are also referred to as subordinates, or the disadvantaged. For example, in the United States, targets include but are not limited to women, people of color, gay, lesbian, bisexual, and transgender people, people with differing abilities, and the working class and poor. Agents are members of dominant social groups that are privileged by birth or acquisition who—either consciously or unconsciously—exploit or reap benefits over members of the target groups. Agents are also referred to as dominants, or the advantaged. Agents include but are not limited to men, white people, heterosexuals, able-bodied people, and the upper-middle and upper class. Agents may also be referred to as allies if working to address oppression against targets.

Performances of identity relate to how social identity is interconnected with discourse (Goffman, 1981; also see Johnstone, 2002). This humanistic perspective explores the ways in which verbal language is a representation or a performance of a person's social identity. In this chapter, comments about one's own social identity, or the social identity of someone else, were coded. For example, when participants explicitly mentioned their own race, the race of another person (present in the room or not), and/or discussed race in relation to an idea or concept, the comment was coded. In the next chapter, I explore performances of identity in more depth—and from a more humanistic perspective.

Johnstone (2002) describes *persuasion* as one of the major purposes of discourse designed for strategic purposes. There exist four types of persuasion: logical syllogism, quasilogical, presentational, and stories. Logical syllogism is a central argument that has a major premise $(x + y = z)$, a minor premise $(z = a)$, and a logical conclusion $(x + y = a)$. Quasilogical persuasion includes arguments that draw upon the logic found in debate or mathematics and yet are not, in fact, logical. An example of quasilogic is a relational argument; if person A likes person B, and person B likes person C, then person A will like person C. However, relationships do not translate in the same way as mathematical formulas; just because person B likes person C, it does not necessarily mean that person A will like person C. The third method used is presentational, where persuasion comes from being personally moved, inspired, and/or convinced through a rhythmic flow of words and sounds such as is found in poetry and some forms of ministering or preaching. This final strategy uses stories in order to create analogies between the context of the story and the current situation. For example, a keynote address at a scholarship award ceremony at the college's center for women may include a personal story about the difficult struggles a woman faced when attending college and the perseverance she needed to graduate. This may inspire the audience and persuade them to continue to support the women's center.

Power (asymmetrical relationships) and *solidarity* (symmetrical relationships) are always operating between and among people (Tannen, 1994). *Institutional power* comes from the operation of laws, policies, procedures, and historical and cultural paradigms. Foucault (1976) argues that whoever holds power regarding what counts as knowledge also has power over policies and systems. For example, tenure and promotion committees have institutional power to determine the policies, procedures, and requirements for a faculty member to obtain tenure. As decisions and policies are made in a relevant historical and cultural context, the target and/or agent identities of a faculty member and the members of the tenure and promotion committee interconnect with various decisions about procedures and policies. More specifically, the tenure and promotion committees at some institutions have the institutional power to determine if additional time may be granted to a person seeking tenure (regardless of gender) when adding a baby to the family. From this perspective, no decision regarding policies and procedures may be "untangled" from historical and cultural contexts related to social identities (gender, sexual orientation, etc.) of the participants.

Situational power describes specific places or situations where people compete for power. For example, if a faculty member is not recommended for tenure, students may protest by sitting in the dean's office, or the department faculty may write letters in support of the faculty member in order to try to gain situational power.

Solidarity is defined as the counterpart of power where people join together in solidarity with people who hold common social identities or roles. A person may state, for example, "As another woman, I also believe..." or "I am also from the Shinnecock tribe and can relate to..." or "I, too, received my degree from a community college and have faced..." as a way of showing solidarity with another person from the same identity group.

Power and solidarity may also be explored through the concept of *footing* (Goffman, 1981). "A change in footing implies a change in the alignment we take up to ourselves and the others present as expressed in the way we manage the production or reception of an utterance" (p. 128). For example, when a person says, "I'm *just* a community organizer...," or shares her opinion and follows it up with "but, I'm *only* a student," she changes her own footing in the conversation to a one-down position. The reverse changes a person's footing to a one-up position. This one-up happens, for example, when a person says, "As a full professor, I ..." or "As someone who works in Washington DC, I know that ..."

Rules of politeness are the rules or guiding principles for the manners in which speakers act (or do not act) in a way that matches the prevailing (dominant) etiquette of the situation. The etiquette of the situation is different between cultures and sub-cultures. For example, the etiquette for participants attending the annual drag show in the Risley Dining Hall at Cornell University is different than the etiquette for the annual gender bender show at the University of Michigan Union. In one case, singing along and dancing in the isles during the performances follows the etiquette for the situation; where as in the other case, most participants sit in seats lined in rows and respectfully watch the performer. However, in either situation, enter the university president and trustees, and the etiquette for the situation may change (or the etiquette President Coleman used at dinner may change from how she expresses herself at the performance).

Lakoff (1973) shares three rules of politeness:

1. Formality (distance)—where a person does not impose him/herself on others,
2. Hesitancy (deference)—where the speaker allows the addressee options about whether to respond to the speaker and how, and
3. Equality (camaraderie)—where the speaker acts as if the addressee and the speaker are equal.

Silence is another form that conversation between leaders takes. The intermittent use of silence by a person when the person has the floor (dramatic pauses and/or silence between words) and the choice not to engage in the discussion by remaining completely silent are ways in which a person communicates. Silence can be used to give the speaker time to think, to hurt others as a response to personal anxiety (e.g., giving people the "silent treatment"), to prevent communication of certain messages, to communicate ideas, or to express that one has nothing to say (DeVito, 1992). At an admissions committee meeting, if a usually vocal member of the committee does not make a comment or share thoughts on an issue, the person's silence relays a definite message to the other committee members.

Silence can be considered in another way. Tannen (1993) points out, silencing a person is not necessarily connected to volubility; a person may talk a lot or use many words to describe a concept and still be silenced. For example, a person may speak up in a higher education committee meeting and may not have felt silenced in the moment, but if hir suggested comments and ideas are not incorporated in revised ideas or committee practices, then the person may feel silenced, particularly if the rationale for not incorporating these ideas is not provided. In this sense, to be silenced in the committee meeting is distinct from silence (or the lack of verbal discussion) during a meeting.

Turns are units of talk when one person starts to speak, does so, and then ends his speech. When a second person speaks, that signifies the next turn (Johnstone, 2002). Turns are structural in nature. A *move* is the action that is taken during the turn, such as asking a question (move) (Goffman, 1981). Moves are functional in nature. The moves, combined together, are known as an exchange between people.

What I term *verbal supportive gestures*, a form of DiVito's (1992) verbal confirmation, are ways in which a listener lets a speaker know that s/he is supportive of what is being said. Verbal supportive gestures come in many forms, including "yes," "um, hmm," "I agree," "that's right," and many more. For example, when a graduate student is giving a speech in the middle of campus to support affirmative action, the "yes" and "um hmm" comments that are shouted out during the speech from the audience are known as verbal supportive gestures.

DESCRIPTION OF CONFERENCE III'S PLENARY AND DISCUSSION: SESSION 3

The plenary and discussions in Sessions 1–4 during this conference series were extremely similar to each other. Plenary Session 3 in the third conference was chosen randomly (through a roll of dice) for further exploration. The session began with two invited speakers delivering a speech on the predetermined topic, "Creating Significant Change in the Campus Culture." The speeches lasted for 5–10 minutes. At the conclusion of the speeches, a facilitator opened up the floor to participants for questions or comments.

Participants directed questions to the speakers and, at times, shared a new idea or a personal/professional story to accompany their question. Many of these stories offer examples of connections between communities and universities, examples that should be studied in more detail in future analyses. Rarely did a participant ask a question of another participant or challenge the speakers or disagree with them. The speakers answered each question with information about their institution/program or by expanding on the idea they had presented. Ideas were often expanded through personal/ professional stories or through persuasion methods of logical syllogism $(x + y = z; z = a; x + y = a)$. When a logical syllogism was used, it functioned as justification for why their institution/program follows procedures or has developed a program in a certain way, or why the speaker believes as s/he believes.

The various turns and primary performing moves in Session 3 are provided in the previous chapter. The breakdown of the various tools for analysis within each move as defined in the literature (various hedging, rules of politeness, questions, etc.) has been compiled for Session 3 and is represented in table 4.2. Coding for tools for analysis began after the invited speakers concluded their speeches and the facilitator invited participants into the discussion. If a specific move or statement fulfilled principles of more than one tool for analysis, then it was coded in both places.

In Session 3, nineteen different people spoke in a total of 42 performance moves. Each person spoke an average of 2.21 times, with the first invited speaker talking most often. Given the definitions for the tools for analysis described above, there appeared to be no elements of code-switching, face threatening acts, or uses of silence during a persons' comment. The number of times a person used "hedging" verbal behavior was further broken down by gender where men appeared to engage in it 23 times and women 15 times, which is not statistically significant when the number of male and female speakers are considered.

The majority of interactions happened when participants asked a question of a speaker (10) or participants aligned with what a speaker had stated (7). At two different points, participants posed questions to other participants in the room. Participants shared approximately eighteen stories about their own experiences at their institution or national organization in order to communicate an idea. In some stories, the participant was a part of the example, in others, participants shared stories that they had heard. For example, one person shared her own experiences participating in a service-learning program that connected suburban students with urban children and discussed how this was relevant to the goals of the conference.

At three different points, the first invited speaker or participants used principles of logical syllogism in order to articulate their perspective. The entire facilitated conversation seemed to follow formal rules of politeness as rarely did people interrupt each other, talk when others were talking, or speak without being called upon by the facilitator. There seemed to be three instances of formal politeness between

Table 4.2   Coding for Tools for Analysis of Conference III, Sessions 3 and 5

Tools of Analysis	Number of Occurrences	
	Session 3	Session 5
Code Switching	0	3
Face Threatening Acts		
Positive face	0	3
Negative face	0	10
A paring of positive and negative face	0	2
Hedging		
Men	23	27
Women	15	14
Graduate Students	0	4
Humor	10	7
Interactions		
Facilitator		
Aligns with participants	0	1
Asks question of participants	2	1
Participant		
Agree/align with speaker 1 or 2	6	—
Agree/align with speaker 1	—	9
Agree/align with speaker 2	—	2
Agree/align with speaker 1 & 2	1	1
Disagree with speakers	0	7
Align with speaker 2, then disagree with speaker 2	0	2
Repeat what speaker stated, adds new idea	2	5
Asks question of speaker 1	5	0
Asks question of speaker 2	4	1
Asks question of speakers 1 & 2	1	0
Agree with speaker earlier in Conference III	1	1
Disagree with speaker earlier in Conference III	1	0
Agree/align with other participant/s in room	0	9
Disagree with other participant/s in room	0	2
Asks question of facilitator	0	1
Asks question of organizer	0	1
Asks question of other participants in room	2	5
Speaker		
Agree/align with participant/s	2	1
Disagree with participant/s	0	4
Repeat what participant stated, adds new idea	0	1
Asks participant/s a question	0	4
Agree/align with facilitator	0	1
Organizer		
Agree/align with speaker	0	1
Disagree with speaker	0	1

Continued

Table 4.2 Continued

Tools of Analysis	Number of Occurrences	
	Session 3	Session 5
Asks question of speaker	2	0
Shapes the agenda/direction of the conference	1	2
Persuasion		
Logical syllogism (x + y = z; a = z; x + y = a)	3	0
Quasilogical (person a likes b; b likes c; so a will like c)	0	1
Presentational (rhythmic flow, verbal art)	0	6
Use of stories	18	13
Power		
Footing (up or down)	2	5
Institutional power	2	4
Situational power	4	11
Solidarity	0	2
Rules of Politeness		
Equality (camaraderie)	1	1
Formal (distance)	2	1
Hesitance (deference)	0	1
Social Identity		
About own self	0	2
About someone other than own self	5	4
Social identity, the topic of	1	1
Total number of different people who spoke	19	30
Total number of moves (comments)	42	86

participants that were particularly apparent—one instance of equality or camaraderie and two instances of formality or distance.

In Session 3, there appeared to be no performance moves when participants shared their own social identities around age, class, race, gender, or such others. However, five different times participants offered the social identity of people other than themselves; in each case this was the identity of a person in a story that they were articulating. For example, in one case, an organizer asked the first invited speaker to talk about his speech in relation to racism, sexism, and homophobia at the institution where he worked. In response, the speaker shared a short answer and the next person asked a question unrelated to issues of race, gender, and sexual orientation.

Once at the beginning of the session and once at the end, the same organizer brought up issues of institutional power (and actually used the word "power"). In each separate case, he asked the first speaker questions about institutional power and how it plays out at the speaker's institution. In both cases the speaker answered the question with a logical syllogism or a story about the institution; other participants in the room did not follow up on the question with subsequent performance moves.

Situational power was observed in four different instances. Three of the instances have already been mentioned—where questions about racism/sexism/sexual orientation and institutional power were raised by the same organizer (an African American man), answered briefly with a logical syllogism or story, but not followed up in more depth by the speaker (also an African American man) or participants. At these times, the speaker and subsequent participants had the opportunity to voice their perspective or ask additional questions related to power and social identity. However, the group moved past the organizer's questions without following up with queries seeking deeper understanding of the topic. It appears that participants used their situational power to address topics other than those related to social identity or institutional power and adhered to formal rules of politeness. In addition, the invited speaker, who quickly answered each question, did not call upon others in the room to contribute additional comments on the topic. These moves relate to whose ideas get picked up and whose ideas do not get furthered in a conversation. In this instance, each person who speaks contributes to the path of the discussion and uses hir own situational power to direct the discussion.

Another use of situational power was observed when the white, male facilitator asked an early career, white, female invited speaker to expand upon a topic. The facilitator asked for specific descriptive and personal perspectives from this woman but did not ask the male invited speaker (or anyone else in this session) for such personal experiences and observations. The facilitator seems to have used his role and power as "facilitator" in order to probe into the personal feelings of this second speaker.

Description of Conference III's Plenary and Discussion Session 5

Plenary and Session 5 was distinctly different from Session 3. The participants focused on strategic alliances through the creation of teams and on various models offered by the organizers and introduced by participants (e.g., institutional models, systems models, change models). In this session, 30 different people spoke in a total of 86 moves. Each person spoke an average of 2.87 times, some more than three times, some only once, and others not at all. An overview of the various turns and primary performing moves found in Session 5 are provided in the previous chapter. The specific breakdown of the various tools for analysis within each move during Session 5 is given in table 4.2, alongside tools for analysis used in Session 3 for a comparative analysis. Coding for tools for analysis began after the invited speakers concluded their speeches and participants were invited into the discussion. If a specific move or statement fulfilled principles of more than one tool for analysis, then it was coded in both places.

There were three different times when two different people engaged in verbal code-switching as defined by Gumperz (1982); both participants self-identified as African American and male[2]. One African American man served as the second invited speaker and the other, a participant. Code-switching happened toward the end of the discussion, when there were more elements of disagreement among people than those of alignment. For example, the speaker used code-switching when addressing questions and expressions of disagreement from a white woman. He then reflected on this discourse and stated, "Our [African American] culture is one where we do, where we jump in on each other and aren't shamed of that. I'm just going to state my business." In this way, the speaker was cognizant of the differences between his style of speech and that of the white woman.

There appeared to be at least 15 face threatening acts in this session. One prominent negative FTA happened when a male participant commented to the person next to him about the second speaker, while the speaker was talking. The male participant said, "He's lost it." This comment was loud enough for the tape recorder in the middle of the room to pick it up. The speaker did not stop talking and may—or may not—have heard the comment.

In Session 5, participants asked questions of a speaker, aligned themselves with what the speaker/s (one speaker or both speakers) presented, or challenged what one of the speakers presented. In addition, many participants spoke directly to each other and when they talked, they did more than simply share personal/professional success stories. The content of the discussion was layered (for analysis of the content, see the next chapter). For example, there were 13 questions asked to individuals or the group by the participants, speakers, organizers, or facilitators. There were 26 comments where people aligned themselves with each other (without then disagreeing with someone in the same sentence). For example, one woman stated, "And your comment about paying attention to graduate students, I think, is a critical one." In this statement, the participant agreed with another participant in order to give strength to her point and align with a person who made an earlier comment.

There were at least 14 different times when people disagreed or did not align with others in the room (without also agreeing in the same sentence). The following interaction between the second speaker and a woman participant is an example of disagreement expressed using a number of different tools for analysis. Italics are used to indicate the speakers' original emphasis.

> Speaker 2:...Nobody speaks to it [access to higher education] with any power, nobody stands up. No, we go ahead and just raise up our trite—raising it up. Where is that movement? Where is it?
> Person 26: What do you mean no one speaks to it?
> Speaker 2: *Well, where is it?!* Where is the public demand for it?
> Person 26: [She laughs.] Some of us do it all the time.
> Speaker 2: Where is it standing at? I mean, I'm an [community] organizer, I know—I know what it takes to create something public.
> Person 26: That's not true.
> Speaker 2: Well, you show me where it is.

In this example, the first comment was coded as "disagree with a participant" and "negative face" as the speaker was addressing a point from earlier in the conversation. The comments from person 26 were coded (once) as a "disagreement with speaker" and "negative face" because—with her comments—she not only disagreed with him but also challenged his perspective. Further, it was coded once as "situational power", when the speaker is challenging her with his comment "Well, where is it?!" and she is challenging him through her disagreement and laughing out loud. In this situation, the laughter (not coded as humor) could have been used potentially to discount his comment or to save face herself. In addition, the "I'm an organizer" comment was coded as "footing." With this comment, speaker 2 positions himself as an organizer to gain credibility indicating that, as a community organizer, he is in a better position than academics to determine whether people are demanding access to higher education.

In Session 5, the participants also used a number of stories about themselves, or their institution, to communicate thoughts or share their successes/challenges. In two cases, participants shared their own social identity. For example, in order to align with the second speaker, one participant stated, "I'm a Black man, you're a Black man. We've got something in common right there, and we've got a lot of other things that we can find in common." This person shares his own social identity groups around race and gender and articulates the already self-defined race and gender of another person. At other times during the session, people shared the social identity of someone not in the room. An example of someone talking about "the topic of social identity," in general (as opposed to naming her own or another person's social identity), was an African American woman sharing a story about her regional state university. She stated,

> We have populations that are very important to us as Americans who are not at the table in the numbers we need them, particularly Chicano and Latina, one of the fastest growing populations in our nation...GLBT [gay, lesbian, bisexual, and transgender] populations are another great issue that we haven't talked about. Most of the young—forty percent of young people on the street—uh, are children from homes who've been kicked out because they were GLBT. Nationally. That's a crazy statistic.

ANALYSIS OF THE SIMILARITIES AND DIFFERENCES BETWEEN SESSION 3 AND SESSION 5

A number of similarities between the sessions exist. The same people were included in Session 3 and Session 5, and the facilitator was the same white man. Each session starts with a speech from a white woman and an African American man; in Session 3 the African American man speaks first and in Session 5, the white woman speaks first. The number of times that men and women hedge at the beginning, middle, or end of their comment is approximately the same. In addition, the number of times participants use humor or share stories of their experiences is approximately the same. Further, the types of stories that people share are relatively similar. However, interactions that happened during the "process" of the conversation—or more specifically, the interpersonal dynamics in each of the sessions—are extremely different.

In this section, I explore the most salient differences between the sessions including those related to interactions between people (alignment, disagreement, questions) and social identity as interconnected with power. It is these differences that led me to consider Session 5 the most dynamic "peak" session and concentrate on this session in subsequent chapters. It should also be noted that the similarities between the sessions are also important and should be explored in more detail.

Interactions between People

In Session 5, most questions and comments are directed from participant to speaker, or speaker to participants; similar to the back-and-forth volleying in a game of tennis. The speakers present their ideas, participants (and one organizer) ask questions of the speakers, and the speakers answer the questions. In Session 5, participants engage with each other and with speaker 2 in particular. Specifically, the participants not only direct questions to the speakers but also ask questions of each other, align with each other, disagree with each other, or have a discussion with (more than simply asking questions of) the speaker. This dynamism is indicative of the depth of the conversation in Session 5. For example, participant 17 asks a number of questions to the group members as he reflects on the conversation, which a few participants follow up during the conversation. He states,

> What I *hear* here is an extraordinary richness of experience in different institutions and that, that *always* is an amazing story when you get groups of people—whether it's inside the institution or among the institutions. The *amount of different exciting* things going on is—always [happening]. Yet, at other levels, like, at the state level or at the national level, this [conversation], everything becomes *amorphous* about higher education. I mean, what are we actually doing? And, why can't we, why can't state legislatures, why can't members of congress *understand* how important all this is when you put it together? Where are the aggregating, if I could use that kind of awkward terminology, uh, organizations to do this for us?

With the comment "What I *hear* here is an extraordinary richness of experience in different institutions and that, that *always* is an amazing story when you get groups of people," participant 17 supports the previous stories of other participants' experiences at various institutions. Then, he goes on to shift the conversation by asking questions to the group about why various state and national policymakers cannot understand the richness of the experiences on the institutional level and questions what higher education organizations are doing to address this problem. Participants 18, 21, 22, and 23 (to name a few) follow up on these questions. For example, participant 22 shares a metaphor to describe his perspective about what is being done to create alliances.

> What I detect here is the "Madge the manicurist" principle at work. That's an obscure reference, which most people here aren't old enough to remember, but it was a commercial for, I think it was Palmolive soap. And, a lady comes in to Madge the manicurist and she says, "My nails are terrible what should I do?" And, Madge takes one look at 'em and she says, "Do everything." And, in a sense—that's right, I mean, everybody, everything, everywhere is a potential, strategic alliance.

Participant 23 responds to participant 22 with some concrete alliance examples of which she is aware. In this comment, she provides a regional and a national model, each of which provides support and works toward making changes. She states,

> There's also a [name of foundation] for helping place graduate students in liberal arts colleges...another model that I was thinking of was the [name of foundation] model, [name of foundation] model on diversity. It focused very much on media and on, um, public changes and that first of all, [it] helped train faculty and administration [on] how to effectively work with the media to get the message they wanted across instead of the *bad* messages, the kind of thing, you know, the kinds of *bad* incidents that happened on campus or was on the news. But the *good* work never did [get shown on the news]. And so they did some very specific kinds of things to—to get the *good* work out there. And then held a series of forums that *included* public changers, as well as faculty and administrators.

In this response, participant 23 follows up on participant 17's original comment and adds concrete examples to further the conversation. This is a strong example of where participants deviate from the volleying of the formal agenda and engage in dialogue with each other.

In Session 5, participants also engage with speaker 2 much more than with any of the other speakers throughout the three days. They engage with speaker 2 by aligning with or disagreeing with his comments. The interaction between speaker 2 and participant 26 mentioned earlier is one such example. In addition, speaker 2 asks questions of the participants during the discussion. For example, when participant 6 asks the group, "How do we empower multicultural frames in non-traditional ways, if we are going to make all of this work?" and mentions that GLBT people should be included, speaker 2 directly asks for clarification regarding what "GLBT" stands for. Participant 2 (along with others) answers that GLBT stands for gay, lesbian, bisexual, and transgender people. In this example, speaker 2 is comfortable enough with the group to ask a question regarding acronyms that he does not know. In addition, he reinforces the "in" versus "out" status of the higher education participants versus participants such as himself, a community organizer. With his question, he makes people aware that the acronyms they are using may be familiar to some in higher education but are not to him. In this sense, he encourages that the definitions or the full descriptions of people or organizations be used during the session.

Simultaneously, he signals that he is not aware of this social identity group as this acronym is not familiar to him. In addition, a number of questions from the participants or speaker 2 are not directly posed to an individual per se yet challenge the entire group to consider alternative perspectives. For example, participant 25 states,

> We talk about social movement here as a model for change. And change is certainly a part of the social movement, but so is risk. And I feel like one of the things we've done in the conversation is talk about changing things but in very safe ways...we still have to have conversations about what we're willing to give up. What sacred cows are we willing to slaughter? What pain are we willing to endure?

In this example, the participant is asking more than rhetorical questions to the group. She is challenging the group to address her questions about what they are willing to give up in order to actualize their discourse about a social movement as a model for change. The questions also provide participants with her perspective regarding what it will take to make change; it will not be easy, and there are tangible sacrifices that the people in the room need to make in order to participate in the movement.

Overall, the interactions between participants and speakers in Session 5 are more dynamic and in-depth than in the other plenary and discussion sessions. Participants challenge each other to cognitively consider the relationships between higher education and society in multiple ways. The content of this passionate discussion is particularly layered and is analyzed in more detail in the subsequent chapters.

Social Identity as Interconnected with Power

In the discussions of Sessions 3 and 5, power (solidarity, institutional power, situational power, a change in footing) and social identity (age, gender, race, etc.) appear to be inextricably connected through participants' discourse. For example, in Session 5, the majority of times that (1) a performance move about power or (2) the topic of power was discussed, it was accompanied by a performance move around social identity. In this section, I limit my analysis to the connections between "performance moves" about power and social identity. In the next chapter, I consider the "topic" of power (what people said about power) as interconnected with social identity.

Footing

As described earlier, one way in which people give or obtain power is by changing footing, either through one-up or one-down performance moves. In one instance, an early career scholar (interconnected with her social identity around age and position) prefaces her comment with, "Um, I just wanted to say [a] couple—we've been talking about social movements and some of these side movements. I mean, I still have a lot to learn, but..." This Latina graduate student puts herself in a one-down position by beginning her comment with the statement, "I still have a lot to learn, but ..." With this move, she reduces the value of her contribution. In another example, a full professor positions himself one-up by listing his past positions and naming himself as an activist in this context. He states, "As director of graduate students at two universities since practically my whole career. Even [when I was] an assistant professor I think I attributed it to the fact that I was such a pain-in-the-ass activist graduate student." With these comments, he gives himself credibility—or power around knowledge and information—as related to his age and experience in various positions. He also positions himself as an activist in this conversation and then urges for national "preparing future faculty" type of programs that give a "strategic map of the kinds of choices" for graduate students. By listing his credentials prior to stating his recommendation, this white man tries to bolster his own credibility through one-up footing.

Toward the end of the dialogue the interaction seemed particularly intense. There was a disagreement (described earlier) between speaker 2 and participant 26 regarding where the movement around access in higher education is located. The following excerpt demonstrates that speaker 2 was specifically talking about people in higher education only writing papers and doing reports but not using their influence or power in order to create a demand for access to education.

> Speaker 2: But, its not there. And, you can talk about D.C. and what not. I know about D.C. very well, too. But there is no national movement around this question that is speaking to and risking and saying that we will withhold. We will hold things in order for this to be done. That's not the case. People speak and *write papers* and do reports, but there is no demand, there is no challenge, there is no *risking* of how.
> Participant 26: That's not true. That's not true.
> Speaker 2: Well, you show me where it is.
> Participant 26: Well, I'll, later, I'm, I'm not interrupting.

Speaker 2: Fine! I don't see it nationally. I don't see in the places I'm at. And I organize—and our network is in every state in the country.

Speaker 2 challenges the notion that there is a national movement around strengthening the relationships between higher education and society. He calls attention to the importance of leverage in order to make social change. In order to challenge this notion effectively, he *positions himself* as someone who is knowledgeable about national politics (in Washington, DC) and as a community organizer who works throughout the entire country. Participant 26 disagrees with speaker 2 and firmly states, "That's not true," articulating that there is demand that each foster a movement, the challenge and risk notwithstanding. When speaker 2 asks for evidence of this, participant 26 uses a politeness move that defers to speaker 2; she states that she will talk about this later and that she does not want to interrupt him.

Solidarity
Shortly after the interaction above, participant 21 identifies with speaker 2 around race and gender showing solidarity with him in the face of the recent disagreement. Participant 21 tells speaker 2,

> I don't want you [community organizer] to give up on us [academics] and say if, if I can't get you guys all, right now, to do things the way we do in the organizing world, then, you know, cut us off. And, I don't want this we/they dichotomy. I'm a Black man, you're a Black man. We've got something in common right there. And we've got a lot of other things that we can find in common. I agree with almost everything that you've said since I've sat in this room.

At this point, participant 21 challenges speaker 2 to continue with these types of dialogues and not abandon the struggles between the community and the university. Then, he aligns with speaker 2 around both race and gender in a move that shows solidarity by identifying both of them as "Black men." In addition, he directly shows solidarity by verbally aligning with the speaker with the statement "I agree." These moves could be to give some power and credibility back to speaker 2 after such a heated disagreement with participant 26. If this is the case, then with this move, participant 21 publicly shows other participants that speaker 2 does have support for his ideas. Or, this move by participant 21 could be to show solidarity with speaker 2 in an effort to have the latter listen to what he is saying in the hopes that speaker 2 will stay with the "movement." If this is the case, then with this move participant 21 encourages speaker 2 to continue the work of building strategic alliances for system-wide change through community-university partnerships, thereby encouraging one more person to give strength and energy to the movement.

Situational Power as Interconnected with the Roles of Facilitator and Organizer

There appear to be four instances of situational power in Session 3 and eleven in Session 5, each connected to the organizers and facilitators. The assertion that the facilitators and organizers are connected to instances of situational power may seem obvious, but there are clear examples where each intentionally uses hir role to shape the agenda.

Situational Power and the Facilitator
As the facilitator for Sessions 3 and 5, he follows through on traditional facilitator-esque tasks. He opens the floor for questions, keeps track of who wants to speak next, calls on various people to speak, and ends the session, adhering to the agenda. These instances are not coded as situational power, as they are consistent with the role that a facilitator is asked to play in these organizational contexts. However, a few times the facilitator breaks out of the "facilitator" role and interacts with the group in a role similar to that of a participant. At these points, the facilitator did not have to wait his turn like other participants; he jumped in and asked his question. When he chooses to do so, he appears to use his situational power as a facilitator to interject his opinion. These interjections are coded as situational power by the facilitator. For example, the facilitator uses situational power in Sessions 3 and 5 when he asks personal and probing questions of women graduate students and an undergraduate. He asks one women in particular clarifying questions, or more in-depth questions, about her personal opinions. The facilitator does not pose similar questions to any of the mid-career or senior career professionals, men or women. Stated differently, the facilitator appears to use his situational power in cases where he has agent identities (e.g., man, mid-career professional) and the participant has target identities (e.g., woman, early career professional or undergraduate). The facilitator may have known these young women prior to this session (or maybe not) and wanted them to share more about their perspectives. In any case, the facilitator assumes he may ask these types of questions to these particular attendees in this context. This is an example of how situational power

and social identity are interconnected and fluid; there does not seem to be a clear place where situational power begins and ends, or a place where social identity begins and ends.

At times, participants take situational power from the facilitator. For example, in move number 37, a white woman professor of sociology (person 24) asks the group for more models other than the ecological model being used by the organizers. She requests, "And what would be useful to me is if somebody could contrast—what are some of the other models of change, which I sometimes hear, kind of, taken for granted?" After her question, the facilitator calls on person 4 (a white, male, full professor), who then states, "There may be someone else who wants to speak to your [person 24's point]." The facilitator responds, "What topic is this? Is this an important [point]?" In this move, person 4 asserts his own power in order to follow up on the request made by person 24 that was not addressed by the facilitator.

The response by the facilitator in this situation could have a few different interpretations. Perhaps the facilitator was not listening to person 24 and sincerely did not hear the request. Or, perhaps the facilitator may not have felt that her request for alternative models was germane to the conversation (i.e., it was not "important") and he was trying to shape the agenda by intentionally moving past her request without allowing for a response. In yet another option, by questioning whether this request was an "important" point, the facilitator could be questioning the move that person 4 made in redirecting the conversation; asking whether it is really "important" could be a way in which the facilitator undermines person 4's attempt to address participant 21's request. These examples demonstrate the complexities of the facilitator role and each participant role in shaping the discussion.

Situational Power and the Organizers
In move number 20, one of the organizers tries to shape the direction of the conversation by stating, "I just, ah, wanted to lob something back into the middle that—that was, I think, at the heart of this session, which is . . . " The organizer then asks specific questions that direct participants to discuss a particular topic. The majority of participants (of all gender and social identities) do not take this direct route when trying to change the direction toward a specific topic. Instead, participants tend to align their comment with the opinion of one of the speakers or participants who has already spoken. In this move, the organizer uses his situational power as an organizer to direct the agenda toward

a specific topic that he considers "at the heart of this session." Stated another way, participants primarily align with a person, whereas the organizer aligns with the topic that he sees as the "heart of" the session. This difference of approach between organizers and participants has important implications in terms of communication dynamics during dialogues. The role power plays in shaping the dialogue may be of particular relevance to organizers as they craft future conferences and facilitate discussions on complex issues.

In another example of situational power exercised by one of the organizers—the organizer tries to speak at the same time that speaker 2 is talking. Quickly following, speaker 2, the organizer, and the facilitator all speak at once. The facilitator prevails and gives the floor back to speaker 2 by saying, "There is a second point he [speaker 2] wants to make." Speaker 2 finishes his point. In this instance, the facilitator intervenes when the organizer tries to talk at the same time that speaker 2 is speaking. In these moves, the facilitator is given more situational power than the organizer; it takes the facilitator—a person with more situational power than the organizer—to intervene and allow speaker 2 to finish expressing his thought before someone else speaks.

In yet another example, this same organizer tries to shape the discussion. In move 39, participant 4 asks whether someone wants to speak to person 24's request to name "some of the other models of change?" Participant 16, a male organizer, states that he is not going to answer the woman's request at this time.

> Person 16/Organizer: We think that this is the problem of a sort, or a set of problems, or a nest of problems or a complexity that requires an organizing movement or perspective. I suspect your question about what other models are available, is invitational but rhetorical, you didn't want us to elaborate those at this moment, I hope not.
> Person 24: That would be another conference.
> Person 16: Yah, yah, at least another seminar.
> TALKING AT SAME TIME:
> Person 24: Individually, that would be helpful, you might mention one. [She stops talking about when person 16 says "one"].
> Person 16: I'm almost too fired up to deliver one, but, ah, but yes. Yes. We believe that part of the excitement of this work could be found in placing values along side a new way of thinking and working and pushing those two together. That that might help us to sustain interest and engagement because people would be learning a new way of working at the same time that they're focusing on values, which they hold deeply, so—if you wanted insight into some of what preceded this, that was it.

Person 24 did not ask for values, new ways of thinking and working, or insight into some of what preceded this conference (which was discussed at the opening "context" session). Person 24 requested models that are an alternative to the ecological model presented at each conference (see figure 3.3) as the model seems "taken for granted" in the discussion. When it was clear that the organizer was not going to share alternative models, person 24 requests at least one alternative model. In the next move, the organizer does not share an alternative model to what was presented but instead describes the excitement in "placing values along side a new way of thinking and working." The final statement "so—if you wanted insight into some of what preceded this, that was it" redefines person 24's request for an alternative model—to become insight into conversations that led up to the conference series. Such a redefinition by a white organizer of a white woman participant's view is not uncommon and is discussed in more detail in subsequent chapters.

At other times, participants seem to give situational power to the organizers. In the following statement, person 22 gives power to this same organizer.

And [organizers name] I would, and maybe if you would address this in your closing comments: who is the constituency of the [sponsoring organization]?…I'd sort of like to be, I'd like to be a little clearer about that than I am unless the answer is—that we're in the process of trying to figure it out.

By addressing the organizer by name, participant 22 directs the conversation toward the goals and needs of the organization as identified by this particular organizer. Furthermore, he gives added situational power to this organizer, as he directly names him. Simultaneously, he does not address any of the other organizers, which speaks to the perception of power of this organizer versus other organizers in the room. Participant 22 requests that this information be shared during the closing comments in the hopes of shaping a part of the agenda to gain clarity about the constituency and connect the preceding comments to the goals of the sponsoring organization.

Situational Power and the Sponsoring Foundation

One element of situational power that was an *absent presence* in the conversation is that of the foundation that is funding the national organization to hold this conference series. In what ways do the wants and needs of the foundation influence the organizers approach to this series? In some cases, the foundation may have an impact on the initial vision of the conference series, decisions regarding participants and designing agendas, conference materials, what is said during the sessions, and what is included in/excluded from final reports. In other cases, foundations remain hands-off. In addition, foundation officers from the sponsoring foundation and other foundations were present at each conference.

SUMMATIVE STATEMENT AND NEXT STEPS

Plenary Sessions 3 and 5 have a few similarities and numerous differences. For example, the same people were included in Sessions 3 and 5, and the facilitator was the same white man. Yet, the numbers that are reflected by the tools for conversation analysis are extremely different from each other, and the depth of each conversation also appears to be very distinct. These differences highlight the ways in which Session 5 was a more dynamic session and serves as justification for why I chose Session 5 for further exploration.

Session 5 will be used to explore the following orienting research question: How do higher education leaders talk about how to strengthen higher education's responsibilities to society? This research question will help me uncover the cognitive processing models (or frames) found in the face-to-face discourse among higher education leaders during this national conference series. In addition, the theoretical and methodological lens will help me to explore the interconnections between discourse and social identity as expressed during the gathering.

In the next chapter, I consider the content of what is said by participants, organizers, and the facilitator and uncover the cognitive processing models that were employed among participants during face-to-face discussion at this "peak" session. In the analysis, I also relate the verbal cognitive processing models to the models as found in the literature review.

CHAPTER 5

Behind Closed Doors: A Critical Analysis of the Discourse

In the previous chapter, I utilize communication theories and conversation analysis (sequence and structure of discourse) to identify the "peak" discussions between higher education leaders who are most engaged with each other. In the peak sessions, there is more dialogue between participants as compared to "non-peak" sessions that are limited primarily to participants asking questions and the main speaker answering them. More specifically, in the "peak" sessions there is a greater amount of agreement, disagreement, engagement, and heated interaction between the higher education leaders. These types of heated interactions tend to be the conversations that people continue to focus on as they reconstruct the dialogue after the fact. They are also the location of the discussion most worthy of further exploration in order to dissect the complexities of the discourse and deepen our own learning with the goal of connecting knowledge and action.

In this chapter, I dive deeper into this micro-analysis by exploring what goes on behind closed doors during these interactive, "peak" sessions. I utilize critical discourse analysis (Cameron, 2001; Fairclough, 2001; Johnstone, 2002), which is concerned with how discourse is structured and shaped by various social forces, in order to zero in on a dynamic session and dissect intergroup dynamics in "naturally" occurring contexts (Peräkylä, 2005, p. 869). What follows is an illustration of the two major findings where I explore both the "content" of *what* is said (or the overt and obvious meaning of the performance move) and the "process" of *how* it is said (including the impact of the performance move on the situation) (Trenholm & Jensen, 1992).[1] This analytic process of sticking closely to the content of the language and the process of interaction enables the participants themselves to identify the critical issues in higher education for the public good. In addition, I connect each of the findings back to the macro-analysis of the literature in an iterative manner. I explore these complex interactions in the hopes that the findings will be meaningful and help move us forward as we consider critical issues for higher education and the public good.

It is important to note that the "non-peak" sessions—sessions where participants showed no sign of disagreement and shared more narrative stories about their own institutions and instantiations of higher education for the public good—should also be explored in future studies. These best practices identified by policy leaders are also important for us to learn from as we work to make educational change.

OVERVIEW OF FINDINGS

The first finding relates to the models presented by conference organizers in earlier sessions, in handout material, or on large posters placed around the room; participants challenged the content of these models. Specifically, participants called to question the value of these models, asked for additional models, or offered alternative models. These challenges to the foundational models primarily came from women and people of color; however, some white men also challenged the organizers' models. Following such a demand, other participants or organizers added to this challenge, rejected this challenge, or did not take up the challenge in any way. The challenges to the content of the model, and the process by which the group did and did not discuss these challenges, mirror the findings in the pilot study[2] (Pasque & Rex, 2010). The challenges to the organizers' models are not incorporated into revised models for change and the models remain unaltered at subsequent conference sessions and in conference summary reports.

The second primary finding through the discourse analysis is that participants of color challenge the actual content of the dialogue engaged in by participants and organizers. In particular, a few of the participants of color argue that a large group of organizers and participants are perpetuating the status quo around educational inequity and are not talking about making substantial change to higher education for the public good. The content of language used by these participants echo the *Advocacy* frame as described in the literature review and analysis. These challenges to the dominant narrative are not taken up or are directly rejected by white participants/organizers. Stated another way, some of the participants and organizers make no room for these particular discourses of advocacy and change around educational inequity. When room is made for challenges to the dominant discourse, it is primarily participants and organizers of color who engage in these types of bridging and inclusive moves.

This chapter does an in-depth exploration of these two challenges as each challenge poses a critical dilemma for the group as a whole when some higher education leaders do not agree. In each case, the dilemmas are not captured in recrafted models or in revised visions for change that hope to strengthen the relationships between higher education and society. Instead, the dilemma is couched in "broad, if not universal, agreement" in the final reporting mechanisms of the conference.

In analyzing this particular discourse, I find something that supports existing literature and something that is quite original. What echoes the existing literature is that women, people of color, graduate students, and people outside the academy are silenced and/or their perspectives rejected (Also see Chase, 2005; Gilligan, 1982; 1987; 1988; Green & Trent, 2005; Rowley, 2000; Smith, 2004; Stanley, 2006; Tannen, 1993; 1994). I use the word "silenced" in this context to mean that ideas are shared and not centered in the discussion and/ or not included in final reporting structures or revised models. As Tannen (1993) points out, silencing a person is not necessarily connected to volubility. A person may talk a lot or use many words to describe a concept and still be silenced. In the case of this national gathering, a participant may or may not have felt silenced in the moment, but when their concepts and ideas are not incorporated in revised concepts of higher education's relationships with society, then the concept or idea is stifled, shut down, marginalized, or silenced. In addition, if someone moves on without addressing agreement or disagreement about the idea (i.e., the phrase "sweeping it under the rug"), then a person may feel silenced.

The contents of this discourse mirror those in the literature review analysis; the ideas or dominant cognitive processing models presented during the conference series prevail during the discussion and final reports. Simultaneously, alternative cognitive processing models presented during the "peak" session are devalued, silenced, and/or rejected. Stated another way, advocacy cognitive processing models that directly address grassroots organizing, educational inequity, and social injustice around race, gender, and class are not centered in policy discussions, final reporting structures, or literature on the relationships between higher education and society. In this way, leaders may miss important options for educational change.

In the sections that follow, these two critical findings are discussed in detail and examples are provided using rich, thick description. I provide quotes from the word-for-word transcript for clarity, using pseudonyms for participants.[3] I intentionally switch from participant numbers, used in the previous chapter, to pseudonyms in order to reflect the interpersonal interaction within a social context; the participants used first names with each other throughout the conference series. This change also signals the shift from conversation analysis to critical discourse analysis. I consider each of the challenges through the theoretical and methodological lenses used in this study and offer a reflective analysis after each finding, thus speaking to the implications of the national discourse through both the conference series (micro-analysis) and literature review (macro-analysis).

Challenges to the Organizers' Models

Conference series organizers presented various models and information in earlier sessions, in handout material or on large posters placed around the room in order to provide context for the discussion. See figure 5.1 for the Ecological Impact Model and figure 5.2 for the Dialogic Process Model. Organizers also verbally described these models and the models were referred to throughout the conference. These were a part of the dominant frames or cognitive processing

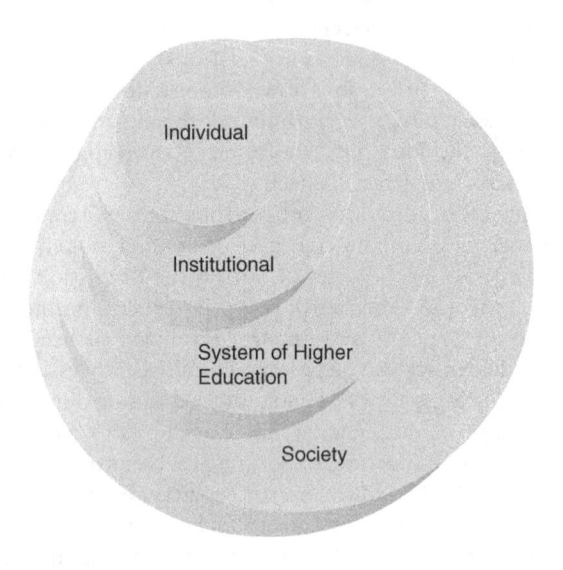

Figure 5.1 Ecological Impact Model

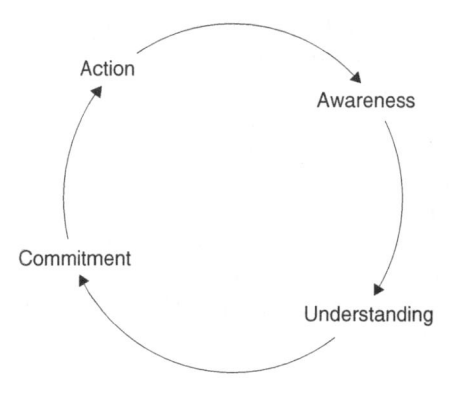

Figure 5.2 Dialogic Process Model

models presented at the conference, which—notably—may be different than dominant cognitive processing models in the literature or at other national policy conferences. A number of participants challenged these foundational models. It was primarily women and people of color who stated such challenges; however, one white male full professor also challenged the dominant perspectives. I share two examples of challenges to the organizers' model and an example of where a challenge was rejected by conference participants. I follow these examples with a short, reflective analysis about the content and process of the dialogic moves.

Description of the Organizers' Models

The ecological impact model is one of the models challenged by participants (figure 5.1). This circular model includes the "individual" level, surrounded concentrically by the "institutional" level, the "system of higher education" level, and the "society" level. Each time adjacent levels overlap, a crescent is formed. The organizers describe the focus of this national conference series as the relationships between higher education and society, or the crescent between the "system of higher education" and "society" levels.

The second figure was the Dialogic Process Model (figure 5.2). This model shows the process of dialogue as a circular movement from "awareness" to "understanding" to "commitment" to "action." The organizers also discussed the model as a principle of the convening, with the goal of the conference series as a movement from awareness to action. Both the standard ecological model and the dialogic model were presented in the material that participants received prior to their arrival, mounted on posters placed on tripods in the room, utilized in Powerpoint presentations, mentioned by organizers in each conference opening session, and referred to by organizers and participants throughout the conference series. The intention or motivation of presenting these models (e.g., to encourage discussion, to use as the basis for encouraging change) or the philosophical basis for these models (e.g., organizational theory, feminist theory) is outside the scope of this research study but could be explored in more detail in the future.

Examples of Challenges to the Organizers' Model

The first time that a model provided by the organizers was overtly challenged during this "peak" conference session[4] was when a white, male, full professor of sociology (Bob) connected the invited speaker's concept of "team" to the ecological impact model. (See performance move #4 in the table "Session 5: Turn Taking and Performance Moves"). Bob states, "And I noticed the psychological [ecological] impact model does not have a *team*, a small group component. So, individuals, to me, are isolated. That's why they are individuals." (Italics are used to signal original participant emphases). With this comment, Bob connects the speaker's concept of "team" and explicitly shows how the "team" concept is absent in one of the primary models utilized by the organizers of the conference.

It is Glenn, the second invited speaker—an African American community organizer—who initially raises the topic of teams as oppositional to "institutional self-interest" during his invited speech and it is Bob who specifically points out that the topic of teams is absent in the ecological model. With his statement, Bob makes certain that participants explicitly understand that Glenn's discussion of "the importance of using teams when creating strategic alliances" is considered in relation to the ecological model; Bob clarifies this connection in order to make this flaw in the model obvious to the participants. This statement signifies that Bob did not believe that this concept was made clearly enough by Glenn, and he needed to stress this point. Bob uses his power as a participant in the dialogue to make certain that the group addresses the concept he feels is important.

One participant follows up on the questioning of the model approximately one-third of the way through the session. In this example, Stephanie, a white woman professor of sociology, returns to the question of the organizing model (performance move #37). She states,

> I think that as a relative newcomer to *this* level of meeting, I have been very excited about what feels like, although I'm not sure, an agreement that the organizing [ecological] model, the social movement model is, maybe, *a* best way to do this. And what, what would be useful to me, and maybe... I'm not thinking that that's always the model that I see that many institutions, I think, including my own, are necessarily using to develop this work. So, I feel that the [organizing organization] in this group could make a contribution if that was part of the shared vision and purpose. And what would be useful to me is if somebody could contrast what are some of the other models of change, which I sometimes hear, kind of, taken for granted. I'm thinking of entrepreneurial models, for example, more, forgive me, top down models for change. Whereas I have tended to agree that the way you brought about change was from the grassroots and organizing, but I don't think that's been taken for granted in the civic engagement [community].

Stephanie starts this comment by putting herself in a one-down position, stating that she is a newcomer to this level of a national meeting. She goes on to articulate her excitement about the "agreement" regarding the ecological model for this conference series. In this move, she articulates how organizers and participants have, up until this session, agreed upon the model. Stephanie mentions that this model may be "a" or one way to think about the topic at hand. The word "a," which she verbally emphasizes, is a key distinction. It signals to listeners that she believes this is one of *many* models that could be used.

Hedging, Stephanie states, "And what, what would be useful to me, and maybe..." The hedging reduces the strength of her request—that the organizers present more than one model for considering this type of work. As a potential way to strengthen her request, she provides personal evidence that her own institution is one of many institutions that does not use this model. She asks that someone contrast "other models of change" that are "taken for granted" in this conversation and that these other models would be useful. This question to the group may be Stephanie's way to get the group to acknowledge the top-down models that many institutions currently use, or to enable the group to think more deeply about the models that are talked about versus models that she perceives are used in higher education.

Stephanie ends her comment by mentioning that she does not think that people involved with civic engagement have always used a "grassroots organizing" model, and that these models are not always "taken for granted." This comment implies that some of her colleagues who do work with civic engagement do not use a grassroots organizing model in their approach to this work.

Women in related conferences about higher education's relationship with society have used similar strategic moves to resist the framing ecological model presented (Pasque & Rex, 2010). I discuss this specific example in more detail in the next chapter.

Rejection of the Challenge to the Organizers Model

The next speaker does not take up Stephanie's request for additional models. At this point, Bob, the same full professor who originally mentioned that the concept of "team" is not included in the ecological model, requests that someone answer Stephanie's question before moving on. In this performance move, Bob views the direction of the conversation as negotiable and steps in to make certain that Stephanie's comment receives a response. This professor uses his power in the room in order to ensure that an organizer respond to Stephanie's challenge to the model.

Bolman and Deal's (2008) political frame is a useful frame to use when considering the resistance moves made by Bob. In the political frame, power is defined as the capacity to enable things to happen. This may be done through influencing behavior or changing the course of events where negotiation and interaction of key players is ongoing. Bob negotiates the ecological model as the standard model, suggesting that alternate models should be considered. Moreover, Bob could be using his situational power as a male ally in order to make sure Stephanie's important point is addressed by the group (for more on allies, see Reason, Broido, Davis, & Evans, 2005). Or, he could view Stephanie's request as one that supports his own resistance to the ecological model (the topic at hand) and help to make certain that the request is entertained as it furthers his own perspective. In any case, his roles and identities provide him with power to influence behavior and change the direction of the discussion in this context.

At this point, the white male facilitator asks, "What topic is this? Is this important?" questioning the importance of Stephanie's request. In response, a white male organizer (Joseph) states, "I suspect your question about what other models are available, is not invitational but rhetorical, you didn't want us to elaborate those at this moment, I hope not." When Stephanie sticks to her initial request and states that she would, in fact, like to hear at least *one* more model, the organizer redefines her original question. Explicitly, Joseph states,

> We believe that part of the excitement of this work could be found in placing values along side a new way of thinking and working and pushing those two together. That—that might help us to sustain interest and engagement because people would be learning a new way of working at the same time that they're focusing on values, which they hold deeply, so—if you wanted insight into some of what preceded this that was it.

In this performance move, Joseph talks about excitement, interest, and engaging. He alludes to a new way of working that is connected with deep values. As a listener, the intonations and content of the words themselves seem quite engaging. Simultaneously, Joseph redefines the question that Stephanie asked by stating, "If you wanted insight into some of what preceded this." Stephanie never requested more information about the preceding conversations that led to the use of this model. She requested different models of change that were not provided during this session. This example connects to research by Gilligan (1982; 1987; 1988) who found that boys/men have a need for an external structure of connection. In addition, boys/men tend to step back from the situation and appeal to reasoning, often losing sight of the needs of others. In this example, Joseph connects himself with others by often using "we" in his response. Importantly, this provides him with an external structure of connection—the group of organizers. In addition, he provides a rationale that is different from what Stephanie requests, thereby losing sight of what she states that she needs. He resists Stephanie's challenge to the models and uses his power as an organizer to reframe her request in a polite manner.

Analysis of Content and Process

The content of the ecological model is a part of the dominant frame or cognitive processing model held by leaders at this conference. The organizers and many of the participants speak favorably about the content of this model. When Glenn discusses the concept of "teams" as essential to "Building Alliances for System-Wide Change" (the title of the session and his speech), it becomes evident that the "teams" he is talking about are not included in the ecological model. Bob highlights this point and, with his performance move, clarifies—what I call—an alternative cognitive processing model. Stephanie also questions the models and asks for other models of change that reflect "grassroots" organizing. The alternative (or non-dominant) cognitive processing models stated by the participants include "teams" and "grassroots" organizing as components for change between higher education and society. Further, the challenges to the organizers' models (by Glenn, Bob, and Stephanie) are not incorporated into revised models for change; the models remain unaltered at subsequent conference sessions and in final reports.

The content of "teams" is also found in some of the literature in the *Pubic Good* frame. For example, Chancellor Cantor (2003) connects departments across the university with a new center for democracy and the expansion of the intergroup dialogue program. In addition, she makes connections between university faculty, staff and students, and community partners—a point that is stressed by Glenn in his speech. The "grassroots" organizing concept is echoed in the *Interconnected and Advocacy* frame through the continual urging of leaders to organize against economic neoliberalism and the marketization of higher education. Through Stephanie's performance move, it is not necessarily clear whether she also echoes these tenets of the *Advocacy* frame, but it is clear that she questions the content of the current model and requests alternative models.

The process of how the ecological model is challenged and how this challenge is (is not) addressed is also worth noting. The content of the ecological model is challenged during this conference session and in earlier conferences (see the next chapter) and yet this same model is used in the final report for this conference series. The model certainly did spark conversation, which is what the organizers had hoped. Notably, the organizers do not offer an altered or revised model. In addition, there was no time set aside during the conference series to incorporate alternative concepts into the model. The final conference series report does briefly mention the resistance of the model but in a short and strategic manner. The document states, "The question, perhaps, is not which model is the right one but rather how to find effective leverage points for institutional change." This language suggests a shift of focus by the organizers from the ecological model to various leverage points for "institutional" change incorporated within the model (on the crescent between the "system of higher education" and "society" sections of the model); however, this focus on the crescent was originally presented during the conference series as a foundational principle. The crescent starts off as and remains the focus of the ecological model during this conference series. With this rhetorical move in the final report, organizers continue to ignore the alternative concept of teams and disregard requests for additional models of grassroots change. In this manner, the dominant cognitive processing models continue to be perpetuated during the conference, throughout organizational processes, and in the formal documentation of the conference series.

As mentioned, the organizers did not leave the conversation here but continued to work through other venues to try strengthening the relationships between higher education and society, both with each other and independently. While some people who do continued work utilize this model, others introduce new models for change that have been informed by this conference series.

As is also found in the literature review, alternative cognitive processes are voiced during the "peak" session but are not centered in policy discussions or incorporated in final reports. Acknowledgment and inclusion of various cognitive processing models—or frames—of the relationships between higher education and society is paramount as leaders make more informed choices about how to work toward systemic and equitable change in education. Understanding multiple models—the ecological model, the dialogic model, and revised models based on alternative cognitive processing models presented—provides multiple frames for considering the relationships between higher education and society. Having multiple frames to choose from becomes important as leaders strengthen arguments for effective policy change, particularly for developing change strategies that address educational inequities around race, gender, and class. If organizers and/or participants dismiss alternative cognitive processing models offered by participants and pay attention only to the dominant frames of understanding, then they will not gain an inclusive understanding of the problem and lose out on developing/knowing additional educational change strategies for social justice.

On a related point, Kezar (2004) argues that if legislators, policymakers, and the public are unclear about why higher education is important to society, then other public policy priorities may gain support at the expense of higher education. For example, the current economic crisis provides an opportunity for policymakers to focus on issues outside of education. With the collective findings of the macro- and micro-analysis, I extend Kezar's statement. If legislators, policymakers, higher education leaders, and the public are exposed only to dominant cognitive processing models, then they will not be able to critically consider multiple strategies to address problems and foster social change that addresses the current and persisting inequities in the educational system. Multiple models for change are important in this context.

New questions, therefore, emerge: How do organizers offer conference sessions on this topic in such a way that diverse and multiple perspectives are encouraged, voiced, supported, and included? How do organizers and policymakers center alternative cognitive processing models, such as the *Advocacy* frame, during national policy discussions? In addition, how do participants with alternative cognitive processing models ensure that their voices are heard in these types of settings? Is this language that threatens the status quo too extreme and thus the reason why a more tempered approach, such as the *Balanced* cognitive processing model, is more palatable in policy circles? Is there an acceptable way to disagree with a model—or alternative models—and still center it for the group to consider?

Challenges to the Dominant Narrative

In addition to challenging the organizers' models in this "peak" conference session, some participants challenge the actual content of the discussion (among participants and organizers). It is primarily participants of color who challenge the content of the discourse in a way that clearly mirrors the language of the *Advocacy* frame. These challenges to the content—or the dominant narrative during the session—are not taken up or directly rejected by some white organizers and white participants. Stated another way, some organizers and participants do not make room for alternative visions to be incorporated into the conference. Further, two academics of color (one of whom is an organizer) share important bridging moves that connect the dominant and alternative discourses.

In this section, examples of challenges to the dominant conference series narrative are presented. In addition, examples of where these challenges are rejected are described. I follow these examples with a reflective analysis of the content and process of the discussion and connect these findings to the literature review analysis in order to bridge the macro- and micro-analyses. In addition, I share performance moves where participants are inclusive of challenges to the dominant discourse during the session. I also follow these examples up with a content and process analysis that connects to the literature.

Challenges to the Dominant Narrative

In the middle of the conference session, Courtney, a Latina graduate student, challenges the dominant narrative in the session. She states,

> What sacred cows are we willing to slaughter? What pain are we willing to endure? I mean, people who participate in social movement

think about risks all the time, they think about retribution. I mean, they pay with their lives, they pay with their futures, their reputations. And, I don't *really* hear us or see us talking about those things. We were talking about making change in very safe ways that allow us to maintain our status. That allows to maintain our privilege and our comfortability [Participant says, "that's right"]...we *have* to be true and we have to get up and admit we made mistakes. We have to get up and we have to tell the truth about things and we *have* to be willing to give some of our own power and our own privilege up in order to make things better for other people.

Woman participant: Amen.

Courtney challenges the group's (organizers and participants) dominant ideology thus far by stating that participants are talking about making change in the relationship between higher education and society "in very safe ways." In her full narrative, Courtney describes truth telling as naming the existing historical and contemporary inequities (e.g., the history of segregation) together with the structural ways to interrupt these inequities (e.g., diversifying the faculty). Courtney expresses that the group is not telling the truth in terms of naming historical and contemporary inequities and structural ways to interrupt these inequities. With this comment, she advocates for continued efforts that increase sources of capital, as highlighted by Bourdieu (1986), and for participants to name simultaneously the power structures that sustain cyclical oppression, as discussed by Foucault (1976). Courtney's language reflects the *Advocacy* frame presented in the macro-analysis of the literature.

Courtney goes on to say that social movements, such as the movement to strengthen the relationships between higher education and society, are about "fundamentally challenging the status quo"—something that she stresses the participants have not done during this conference series. She articulates that even though they are engaging in conversations, participants in the conference are not addressing what they are willing to give up to make change. Participants are "talking about making change in very safe ways that allow us to maintain our status." Courtney resists the status quo by posing a direct question to participants through her use of the sacred cow metaphor: "What sacred cows are we willing to slaughter?"

As Courtney becomes more assertive in her challenges to participants, she hedges less often and becomes more emphatic with her weight on particular words (as noted with the italics). Her statement is met with verbal supportive gestures from other participants of "that's right" during her statement and an "Amen" at the end of her statement. Courtney's performance move is a point in the session where the conversation topic changes (a horizontal performance move) and participants begin to talk more deeply about the way in which participants are engaging with each other and with the topic of educational in/equity (vertical horizontal moves), but only after a male speaker expands upon her comments. This is different from previous conversations, which focused on higher education's relationships with society in general or with specific best-practices.

Glenn follows up on Courtney's challenge to the group by questioning what organizers and some participants have called a "movement." He states, in part,

> Let me throw out what I am deeply concerned about and what I see as happening inside of our democracy and what's at stake…There is a profound impact taking place on students—young men and women, older men and women who are entering public, pulbic higher education. There is no money for them to go to school, other than borrowing. We are generating a generation of young men and women who are going to come out of schools with debts in six figures [A few participants say "yes"]. That is going to have a profound impact on how they view their lives, how they view their work, and why they come to the institutions that you are in. Why are not folk in higher education standing at the battlements to demand that the access to higher education should be universal and, as much as possible in our culture, free and affordable to people? That is something *that you can speak to*, but nobody speaks to it with any power. Nobody stands up. No, we go ahead and just raise up our trite—raising it up. Where is that movement? Where is it?

Glenn says that issues of financial access are something that conference participants are in a position to address but then states that no one with any power speaks up on this topic—there is no movement. He asks for evidence of the movement by inquiring, "Where is that movement" that participants are talking about? "Where is it?" This language also mirrors the discourse found in the *Advocacy* cognitive processing model.

Rejecting the Challenges to the Dominant Narrative

White people did not appropriate the content of Courtney's words in later conversation but Glenn did use her words. Other than Glenn, subsequent speakers did not refer to her comments even though there was verbal agreement with her statement by people saying "that's right" and "amen." In addition, her comments were not incorporated into the final documents for the conference series. This is one example of where alternative cognitive processing models that challenge the status quo and encourage shared power and privilege are relegated to the margins. Courtney's alternative frame is not included in follow-up processes by participants and organizers supporting dominant processing models, even though she received verbal support from some other participants.

It was the *content* of Glenn's comments that was addressed in this session. White people who spoke during this session tended to partially disagree with or outright reject the content of Glenn's comments. People of color who spoke during this session either agreed with "most" of what Glenn said or reflected on his comments in a way that was supportive (See table 5.1). For example, the first person to resist Glenn's comments was Paige, a white woman college president. This exchange immediately follows Glenn's comments and his question "Where is that movement?"

> Paige: What do you mean no one speaks to it?
> Glenn: *Well, where is it*!? Where is the public demand for it?
> Paige: [she laughs] Some of us do it all of the time.
> Glenn: Where is it standing at? I mean, I'm an organizer, I know, I
> know what it takes to create something public.
> Paige says at the same time Glenn is speaking: That's not true.
> Glenn: But, its not there. And, you can talk about D.C. and what not.
> I know about D.C. very well, too. But there is no national movement around this question that is speaking to and risking and saying that we will withhold. We will hold things in order for this to be done. That's not the case. People speak and *write papers* and do reports, but there is no demand, there is no challenge, there is no *risking* of how.
> Paige: That's not true. That's not true.
> Glenn: Well, you show me where it is.
> Paige: Well, I'll, later, I'm, I'm not interrupting.
> Glenn: Fine! I don't see it nationally. I don't see in the places I'm at.
> And I organize—and our network is in every state in the country.

In this exchange, Paige questions Glenn's assertion that no one speaks to access to education. When he verbally responds with a question asking her to name where it is, Paige laughs. This laughter has the potential to reduce the importance of Glenn's position and/ or could be a way in which Paige responds to tension. Paige shares her perspective on how some of the participants (or people in higher

Table 5.1 Participants Who Address Glenn's Challenges*

Participant	Identity	Performance Move
Paige	White Woman College President	Rejected Glenn's comment
Lamont	African American Male Assistant Professor	Agreed with "most" of what Glenn said
Terrance	African American Male Organizer	Reflected on Glenn's comments and related them to comments made by the CEO of a foundation

* *Note*: Glenn's comments were, in part, a follow-up to Courtney's challenge to the group.

education) constantly speak to the demand for increased college access. Glenn further questions her and asks for a specific example. Glenn states that he is an organizer who is familiar with Washington DC. With this move, he may be trying to position himself with additional credibility—that he is familiar with the topic and with people speaking, writing papers, and doing reports on the topic, and that these same people are not taking risks in order to actualize change. Paige continues to argue that this is not true; when Glenn asks her to show him where it is, she backs down with a politeness move and states that she will not answer because that would be interrupting. Clearly, these two participants are not in agreement about how to address critical issues of higher education for the public good, even though they both believe that the relationship needs to be addressed in some capacity.

A little later in the session, Paige speaks to the discussion that she and Glenn were having earlier in the dialogue.

> What I wanted to say, actually coming after what you [Glenn] said sounds very Pollyanna-ish. Um, because what I wanted to say is that I think an awful lot of good things are happening right across the country... There are lots and lots of good examples of really fine work that is happening and we should focus on the successes and not just on, on the failures and we should feel good about what we do. And some people see life as a, as a fight and as one movement or one group *against* another group. And, other people see life as using your intelligence and your ability to try and do good and make something good happen. And, you don't always do that through a war. Sometimes there is a war and sometimes there is bloodshed. But sometimes you can have a bigger success and make a larger social change simply by a thousand small, positive actions. And nobody gets hurt. And everybody

wins in the end…I think that we can have social change through positive discussion, through everybody participating, and through people working together.

Paige qualifies her comment as sounding very "Pollyanna-ish," which reduces the strength of her argument (Gilligan, 1982; 1987; 1988). However, she follows up by sharing her perspective that good things are happening across the country. Paige urges that social change can happen "simply by a thousand small, positive actions" as opposed to creating a movement where one group is "against" another group. This perspective focuses on the current, positive examples where higher education and society have crafted strong relationships, sets an expectation that everyone should participate and work together, and implies that everyone will win in the end. Paige's perspective is more reminiscent of DesJardins' (2003) "win-win" perspective than Labaree's (1997) "no-win" perspective described in the literature review. Notably, Paige uses the "win-win" principle of the dominant cognitive processing model in the *Private Good* frame to reject Glenn's comments about how higher education leaders are not taking risks to address social inequities. She also uses elements of the Public Good argument by encouraging a thousand small, positive actions as can be found in myriad service-learning initiatives across the country and the globe.

The implication of Paige's statement is that, even given the socio-political context of in/equity, the current trajectory will create a situation where "everybody wins." Contrary to Paige's perspective, scholars from the *Advocacy* frame, such as Giroux and Giroux (2004), argue that higher education leaders do need to "take back" higher education. Rhoades and Slaughter (2004) echo this sentiment and argue that faculty have been too complacent in the face of increase in the marketization of higher education. They argue that higher education leaders need to provide access to postsecondary institutions, prepare citizens for a diverse democracy, and address social problems and issues. The "everybody wins in the end" argument may be precisely what the authors argue needs to change; however, it is the "everybody wins" frame that is included in the dominant cognitive processing models in this national policy conversation.

Analysis of Content and Process

In the opening session, an organizer does speak to the economic and social relationships between higher education and society. The final report mentions this as well. However, the content of the discussions about inequities, structural change, and participant views perpetuating the status quo talked about by Glenn and Courtney are not addressed by most participants in this session, by Paige in her retort, and by the drafters in the final report.[5] The topics of race, gender, poverty, and other critical issues are not fully addressed in this discursive "peak" space but are instead being overtly rejected by participants, such as Paige. These same comments are, however, addressed in important bridging moves described in the next section. The question remains: how can people with alternative cognitive processing models that address issues of educational inequities intervene in a collective discourse in ways that these critical concerns become included in the discussion? Are there ways for people to disagree with the alternative ideas and still center these perspectives, or is the only alternative "to disagree and move on"? What does it mean to agree and move on with the discussion without centering these perspectives?

The findings from this micro-analysis of the conference series discourse mirror the findings from the macro-analysis. Some scholars mention inequities in education. Other scholars address the complexities of the economic and/or social benefits of the relationships between higher education and society—and these perspectives are often relegated to the margins. For example, in the *Advocacy* frame, Labaree (1997) discusses the narrow pursuit of private advantage at the expense of the public. Labaree reflects that the state of higher education is often mirrored through state budget allocations as more of the cost for college is moving away from the state and being placed on the individual family. In the context of the current economic changes, proportions of college costs paid are dramatically changing for stakeholders such as individuals, institutions, states, and foundations. In a related critique, Gildersleeve et al. (in press) argue that higher education policy continues to fall victim to and perpetuate the current era of conservative modernization in the academy; the authors call for critical perspectives to break this cyclical process.

Courtney and Glenn's urgency around speaking up against educational inequity is reminiscent of the alternative cognitive processing model of the *Advocacy* frame. Courtney and Glenn's argument can also be found in Brint and Karabel's (1989) description of the caste system that has been created through universities, community colleges, and trade schools around race, class, and gender. The authors map the disparity between the "Dreams" of American youth to go to college and the diversions that take place within the hierarchy of the system. To reiterate, if a student attends a two-year institution or a four-year regional university, they have little chance of gaining access to resources at a research or Ivy League institution. Furthermore, even when students utilize community college as a stepping-stone to a four-year institution, there are disparities in the type of education students will receive during their college career (also see Giroux, 2001; Green & Trent, 2005; Labaree, 1997). As mentioned in the literature review, rates of college enrollment for African Americans and Latina/os as compared to whites are much lower. The gap between African American and white continuation rates has been widening since 1999 (Mortenson, 2006) and the gap between Latina/o and white rates is even wider.[6] These quantitative differences are further exploited by the qualitative disparities between the various colleges and universities (Labaree, 1997; Galston, 2001).

It is Courtney and Glenn who verbally address these issues of inequity in this high-stake policy conference. They share their concern that access is not equitable and all institutions are not created equal. As Galston (2001) states, "education serves as a sorting mechanism" (p. 225) and increasing the number of graduates does not automatically remove the systemic degradation that is inherent in the academy.

With the content of their discourse, Courtney and Glenn resist the dominant cognitive processing models in the discussion and simultaneously insist that the leaders *in the room* address issues of access in education through speaking up and taking risks. However, Paige urges that higher education leaders are speaking up and the thousand small actions performed do address this inequity.

In a related example from the literature, sociologist Bauman (1999) utilizes the topic of public and private good to consider political theory and democracy. Bauman states,

> Society, to be autonomous, needs autonomous individuals and individuals can be autonomous only in an autonomous society. This circumstance casts doubt on the preoccupation of political theory in general, and the theory of democracy in particular, with the separation between the public and the private domains and their mutual independence. It is rather the link, the mutual dependency, the communication between the two domains which should lie at the centre of both theories. The boundary between the public and the private which these theories are so keen on drawing should be seen as an interface, rather than viewed after the model of the closely guarded inter-state border meant primarily to slow down and limit the cross-border traffic and to sift out illegal travelers. (p. 86–87)

To Courtney and Glenn, the two domains are interconnected and, if viewed as distinct, are sorting mechanisms at best. Giroux and Giroux (2004) echo these sentiments and articulate that public education is being redefined as a private good in order to further stratify the white upper/upper-middle class and the poor/working class who are predominantly people of color. The authors name the inequities of race and class that Courtney urges in her "truth telling" argument. In a similar manner to Courtney, the Girouxs state that higher education administrators and faculty members are passively enabling corporations to take over colleges and universities and that this passivity contributes to the social stratification and perpetuation of inequity. Giroux and Giroux (2004) argue that strengthening the relationship between higher education and society requires "rejecting the model of the separation between the public and the private domains and recognizing instead their mutual dependence" (p. 40). The lines between the historically separated entities are blurred and a university that acts as a public good must recognize that the false dichotomy of public and private has historically perpetuated racism and classism; a university might "open up a space for more than just a democracy" (p. 213).

This challenge of addressing inequity can also be found in the statistics that signal continued stratification and privatization of higher education. A number of scholars from various frames have argued against competition and the privatization of higher education, as they believe it continues to stratify the privileged and the oppressed (Blumenfeld & Raymond, 2000; Bowen & Bok, 1998; Gildersleeve, Kuntz, Pasque & Carducci, in press; Giroux & Giroux, 2004; Green & Trent, 2005; Hagedorn & Tierney, 2002; Labaree, 1997). For example, higher education decreases the likelihood of the habit of cigarette smoking and increases the likelihood of participation in leisure activities and life expectancy (IHEP, 1998). Further, 77 percent of the children of college-educated individuals go to college, whereas only 33 percent of children of individuals without a college education attend college (Gándara, 2002). In light of these figures, the people privileged enough to receive a bachelor's degree continue to live longer, healthier lives than people without a college education, and this is passed down through the generations. Lesley Rex and I refer to this intergenerational progression as *cumulative privilege* and *cumulative oppression* (Pasque & Rex, 2010).

In addition, research shows that students of color and students from a lower socioeconomic status (SES) are less likely than their peers to receive information about financial aid and funding opportunities (Kezar, 2005; Southern Education Foundation, 1995). Kezar (2005) argues that the high price of college education, even if it is subsidized, leaves a "financial aid gap" or "sticker shock" (p. 32), which deters students from attending college or from attending a more expensive institution. Gutmann (1999) contends that a high-tuition, full-scholarship policy for students of color and students of need would promise both equity and efficiency in terms of increasing access to higher education. However, Gutmann states that this may come with its own set of obstacles such as students not feeling as though they belong. She argues that by keeping tuition low, institutions also keep the external barriers to admissions low. Low-income students feel "more like equal members of a low-tuition university than they do with expensive universities where they are among a minority of students who are fully subsidized" (Gutmann, 1999, p. 224). Yet, the price of postsecondary education continues to rise as state allocations for public colleges and universities continue to fall (Brandl & Holdsworth, 2003; Cage, 1991; Hansen, 2004) and this pattern continues in the face of today's dramatically changing economy.

Higher education leaders must not be too quick to dismiss alternative cognitive processing models that directly address educational inequity, access, and social justice, such as the frames offered by Courtney and Glenn. This type of dismissal raises serious concerns as higher education leaders work to strengthen the relationships between higher education and society through community-university partnerships in areas such as health, prisons and the arts, get-out-the-vote initiatives, educational outreach programs, and other community-university partnerships. As programs, initiatives, and policies are developed, they must include perspectives that have not traditionally been centered, otherwise we risk reductive perpetuation of only the familiar and the known.

Inclusion of Challenges to the Dominant Narrative

When challenges to the dominant cognitive processing models at this conference are taken up in a direct and inclusive manner, it is participants or organizers of color who strategically further these ideas. In addition, when people of color from the academy address the challenges, they do so in a way that supports or reflects upon the speakers' ideas. In this way, these strategic bridging moves are inclusive of (1) the person who challenges the dominant narratives and (2) the people who support the dominant narratives. For example, two African American men follow up on Glenn's challenges to the dominant narrative. The first person is a postdoctoral researcher and the second is a part of the conference's organizing team. In the statement that follows, the first participant challenges Glenn to stay in these types of conferences.

> Lamont: Glenn, I'm going to challenge you to not give up on coming to these kinds of meetings-
> Glenn: No.
> Lamont: -because it makes you uncomfortable. I, I might come to one of the kinds of gathering that, that you hold and I might feel totally uncomfortable. As a matter of fact, I want an invitation from you to come to one of, to one of your meetings. But, I don't want you to give up on us and say if, if I can't get you guys all, right now, to do things the way we do in the organizing world, then, you know, cut us off. And, I don't want this we/they dichotomy. I'm a black man, you're a black man. We've got something in common right there, and we've got a lot of other things that we can find in common. I agree with almost everything that you've said since I've sat in this room. But I felt uncomfortable, you know, saying, "Man, how can I approach him?" He thinks I'm part of the academy and, you know, I'm, I mean, I felt uncomfortable. But, I'm going to unburden myself of it because I believe in what you're saying. [Lots of laughter from participants.]

Lamont challenges Glenn to continue to participate in these types of conferences (content), a challenge that indicates that he values the role that Glenn is playing in this particular session (process). Lamont strategically tries to connect with Glenn, validating "most" of what Glenn stated. Lamont aligns in solidarity (Gumperz, 1982) with Glenn in reference to gender and race by saying, "I'm a black man, you're a black man." He continues with that alignment by mentioning additional things they may find in common. This move of solidarity could be a way to save Glenn's "face" in the dialogue. For example, face threatening acts (FTA) are verbal language that pose a threat to a person's positive face (a desire to be approved of) or negative face (a desire to have unimpeded actions) as described in chapter 4. In the session, Paige was threatening Glenn's positive face by rejecting his challenge to the group. Lamont simultaneously signals to other participants in the room that he agrees with most of what Glenn has said and "believes" in what Glenn is saying (content). This provides additional credibility and voice to Glenn's challenge to the higher education leaders in the room (process). By aligning with Glenn, Lamont may be attempting to help Glenn "save face" and/or center this alternative cognitive processing model.

Lamont also rejects the dichotomy between the community organizer and the higher education leader. Lamont simultaneously rejects Glenn's view that he, as an individual in the room and member of the academy, holds power to make change to strengthen the relationship between higher education and society. With this combination of moves, Lamont demonstrates important bridging moves between the dominant and alternative cognitive processing models to ensure that Lamont's challenge is not completely disregarded. In addition, Lamont articulates that he "felt uncomfortable" when Glenn labeled him a member of the "academy"—a member of the higher education community with power to make change. This is when Lamont "unburden[s]" himself of this uncomfortable feeling of being in a position of power and to take risks. Even though his comment is not humorous, Lamont is met with laughter from participants—a laughter that could signal agreement and/or a release of tension during an intense discussion.

In another example, an African American male organizer takes up Glenn's comments. He reflects on Glenn's words and relates them to a speaker at an earlier, related conference. Terrance states,

> The thing, the epiphanal moment came around when, um, when you [Glenn] were saying that nothing is being done and there are no conversations. And, I *know* there, I know there's a lot of folks sacrificing—there's a lot of *stuff* that's of value, when you and [Paige] had the exchange. But, I was reminded of [speaker in a different conference] who is the, ah, CEO, or outgoing CEO and president of [foundation]…she made a comment that higher education on many of these issues had a strange silence. And, it struck me that you were saying basically the same thing. And that our reaction was almost identical that, ya, we do a whole lot of stuff but the people with whom we want to play with and partner with don't know this.

Terrance provides strength to Glenn's argument by relating Glenn's perspectives to an argument made by the president of a foundation. He also reflects that participant responses in the two conferences were very similar. Terrance also mentions what the CEO had said— "that higher education on many of these issues had a strange silence." Evoking the words from another conference about the silence of higher education helps to center Glenn's comments in this setting and provides support that some people (this CEO outside of the current conference) do agree with Glenn.

Instead of agreeing with Glenn himself, that there is a silence on issues of access and equity in education, Terrance shifts the conversation to support Paige's comment that "we do a whole lot of stuff" but people "don't know this." The problem with the relationships between higher education and society then shifts—from what people are doing to make change in terms of access and in/equity to a problem of marketing and sharing of information. In this strategic bridging move of dominant and alternative cognitive processing models, Terrance may be attempting to address the participants' reaction to Glenn's challenge and help Glenn "save face" while not agreeing with Glenn's assessment of the "movement." In this tactical way, Terrance is able to support Glenn while not disagreeing with participants.

The importance of strategic bridging moves by "boundary-crossers" in higher education policy conversations cannot be stressed enough. Nancy Thomas (2004) defines "boundary-crossers" as listeners who connect multiple contrasting views and ideas. Following dialogic and inclusive leadership processes, they focus on long-term goals and relationships. Boundary-crossers are at ease with numerous people, are comfortable in informal or structured conversations, and continually refine perspectives. Such boundary-crossers have the ability to connect disparate frames of critical issues of higher education for the public good in a way that enables all perspectives to be heard.

Analysis of Content and Process

Courtney initiates this topic of discussion and it is Glenn who follows up on the topic. Glenn's comments become more emphatic before participants and organizers verbally address his remarks. The interaction between Courtney and Glenn is reminiscent of a colleague of mine who refers to herself as having to take on the role of an "angry woman of color" in her regular research team meetings[7]. As she describes, it is only when she expresses herself as an "angry woman of color" that her comments are addressed by the white people in her research group. Stated another way, the "content" about educational inequity in research processes is addressed only when the "process" she uses for communication fits in this self-described stereotypic manner around race and gender. This consistent cycle has relegated her voice to the margins and added anxiety to her research experience, to say the least.

In a similar vein, Behling (2007) urges facilitators of intergroup dialogues on the topics of gender, race, class, and religion to be mindful that marginalized viewpoints are not often heard unless conflict is present, otherwise the comments get passed over and people are silenced. Facilitators, he encourages, need to pay attention to these dynamics in order to ensure that all viewpoints are considered. This relates to Chase's (2005) finding that "if a previously silenced narrator is to challenge an audience's assumptions or actions effectively, the audience must be ready to hear the narrator's story—or must be jolted into listening to it" (p. 668).

In light of these examples from the conference session and my colleagues' experiences, my observations, and the audio tapes, I assume that if participants of color (Courtney and Glenn) had not been "passionate" through the nature of their language (e.g., such as intonation, pitch, volume), then fellow participants may not have addressed the content of their comments that challenged the dominant cognitive processing models through naming inequity and the status quo. The comments that urged participants to address educational inequities were certainly marginalized, silenced, and verbally rejected in the discourse of the conference and/or in the primary focus of the final reports. To reiterate, silencing is not necessarily connected to volubility (Tannen, 1993).

These types of challenges to the dominant perspectives offered during the conference series are briefly mentioned in the final report; however, less volatile challenges from other sessions—not those that stemmed from this particular session—are utilized. Can we speculate that if participants who discussed educational inequities—and the relationship of higher education leaders to these educational inequities—did not persist, then they would not have been in the final report at all?

In addition, this calls into question what happens behind closed doors when the reports are written. From my colleagues experience and the citations provided above, could it be possible that behind closed doors it was the organizer of color who worked to include these challenges to the dominant cognitive processing models into the final report—albeit still within the *Public Good*, the *Private Good*, and/or the *Balanced* frame? Why was it that the less volatile challenges were included as opposed to those that addressed issues of inequity and called higher education leaders in the room to be activists for change? What role does the absent presence of the funder play in the shaping of the final reports? Must it always be people with target identities who raise questions and include marginalized viewpoints about educational inequities? Or, must we always rely on allies like Bob, or organizers who use strategic bridging moves like Terrance, in order to help create space for alternative perspectives to be addressed?

DEALING DIRECTLY WITH EDUCATIONAL INEQUITY

Participants who shared alternative cognitive processing models of educational inequity and change simultaneously questioned some of the dominant cognitive processing models of the conference. Many of the people who held dominant frames did not deal directly with questions and comments raised by people with alternative perspectives who spoke of educational inequity or how the leaders in the room could make change. The people who did deal directly and center these alternative models were often organizers/participants of color; in only one case a white male professor highlighted the resistance initially mentioned by a woman and a man of color, and only after he raised awareness were the comments discussed by the group (albeit not completely addressed). Taken together, these examples illustrate how white male discourse is privileged in academic circles and that this discourse does not always deal directly with all of the issues at hand.

More specifically, people who held dominant frames did talk about issues of social justice and educational equity from a *Private Good*, *Public Good*, or *Balanced* perspective, but if the content was that of an *Advocacy frame*, then the comment was rejected or marginalized in the discourse. It is also important to note that this conversation as recorded on tape seemed far more heated, engaged, and dynamic than as expressed on the written page. To use language from a person who served as a member checker, "the emotions are even stronger than you conveyed."

Smith's (2004) concept of racial battle fatigue is useful in uncovering more of the dynamics between people who articulated dominant and alternative frames during this conference session. Smith describes racial battle fatigue as "a response to the distressing mental/emotional conditions that result from facing racism daily (e.g., racial slights, recurrent indignities and irritations, unfair treatments, including contentious classrooms, and potential threats or dangers under tough to violent and even life-threatening conditions)" (p. 180). Smith relates various documented psychological and physiological symptoms (tension headaches; backaches, trembling, and jumpiness; chronic pain; upset stomach; extreme fatigue; constant anxiety and worrying; etc.) to the combat fatigue experienced by military personnel.[8] In addition, the anticipation of a racist event may add to the stressor. Smith furthers that "unfortunately for African American faculty...higher education was and continues to be much more racially exclusive, oppressive, and antagonistic than society at large" (p. 185).

Glenn, a community organizer, and Courtney, a graduate student (e.g., they are not faculty), are participating in a conference series organized by people in the field of higher education with a focus on higher education. White participants and organizers often meet Glenn's questioning with resistance and do not address Courtney's comments in the session or in final reporting processes. In conjunction, it is only with additional verbal emphasis (stating something louder or reiterating it in a different way) that Glenn's comments are addressed. The psychological result of interactions engaged in during this session appears to meet the description of racial battle fatigue. However, it is only Glenn and Courtney who may answer this assertion for certain.

In addition to research about discussions across race in higher education, there is research that addresses the relationship between community organizer/university partners (such as Glenn) and university faculty and staff. Bok (1982) mentions how community members' high hopes are often met with failures that "actually heightened local suspicions and frustrations rather than improving relations with the university" (p. 46). Rowley (2000) furthers this perspective and offers a major premise for considering the relationship between universities and African American urban communities. He states, "The *habitus* of higher education inflicts symbolic violence on individual universities (by defining the traditional university mission and values), and universities in turn inflict symbolic violence on society by either contributing to urban social reproduction (restricted access to and limited distribution of dominant cultural capital) or refusing to help alleviate urban problems (placing little value on public service and civic responsibilities of universities)" (p. 56).

Rowley's analysis may also add insight into Glenn's perspective, his experience during this conference series, and his daily work as a community partner. Glenn may be attempting to make this symbolic violence visible to leaders through his comments in this session. In turn, his persistence and dedication to expressing this alternative cognitive processing model in the company of people who he perceives as having "power" to make change may add to his own racial battle fatigue. Moreover, we cannot leave it only to people of color to address the issues of inequity and marginalization; all higher education leaders could engage in discourse strategies that articulate and center perspectives of in/equity.

It is people of color, women, and one white man who challenge the models of the organizers or the dominant discourse of the participants and organizers. It is a community organizer and a graduate student who directly address issues from the *Advocacy* frame. A postdoctoral researcher of color and organizer of color also address the issue through important bridging moves. What follows is that participants who hold alternative cognitive processing models, directly address issues of educational inequity around race, gender, and class, and challenge the dominant cognitive processing models are being marginalized in the context of this national conference and experience a form of battle fatigue. If this is the case, then participants who continually address issues of educational inequity and hold target social identities (e.g., person of color) and marginalized roles (e.g., community partner) may experience a "double dose" of battle fatigue; one dose for each target identity, target role, or target perspective. This "double dose" reflects the "multiple jeopardy" concept described by Deborah King (1993) in "Multiple Jeopardy: The Context of a Black Feminist Ideology." Further, by not directly addressing issues of educational inequity around race, gender, and class, the dominant cognitive processing models contribute to Rowley's (2000) description of symbolic violence and social reproduction.

Hurtado (2007) emphasizes that "it is time to renew the promise of American higher education in advancing social progress, end America's discomfort with race and social difference, and deal directly with many of the issues of inequality in everyday life" (p. 186). The arguments and research about issues of race, class, and gender spoken from an advocacy frame are present on the margins of the literature and face-to-face discourses. Writings and voices that directly address inequality are not often included in the dominant cognitive processing models presented at high-stakes conferences and continue to be marginalized in powerful circles. If higher education leaders truly want to find ways to "deal directly" with issues of educational inequality in ways that will make change regarding race, class, and gender, then we must foster discussions about inequity, but in such a manner that all voices and perspectives are included. We must center the perspectives on the margins and include voices of participants from target identities. Unlike some participants in this conference series wanting to be "comfortable" (a participant's word) in these types of settings, I believe this *process* of discussing the *content* of race, class, and gender as connected to educational inequity *is* going to be uncomfortable. And, further analysis of the language around critical issues regarding higher education for the public good is vital to our future.

CHAPTER 6

Behind Closed Doors: Women's Voices of Resistance

"Women still do not have equal access to privileged professional discourses or to dominant speaker positions within them…As a consequence, they still struggle to make themselves heard and to have their interests served" (Talbot, 1998, p. 222). In the preceding chapters, I have argued that "hearing" all perspectives provides education leaders with vital options needed to make policy change that may address educational inequities and serve the needs of all people; the absence of particular perspectives or voices reduces options and alternative frameworks with which to consider needed educational change.

Each of these perspectives as well as laws, programs, and educational policies have direct implications for people in society. Oppressive practices and educational inequity have not been extensively interrupted, nor have ways been found to forge stronger relationships between higher education and society in order to do so. Higher education is at a tipping point where new and innovative strategies are needed in order to alter the current trajectory and make needed educational and social change. Yet, how is comprehensive educational change possible in a system that marginalizes perspectives that may be useful, including those from women?

Bensimon and Marshall (2000) state that higher education policy studies "assume academic structures, processes and practices are gender blind. The lack of attention to gender, both as conceptual category and analytical lens, means that the differential experience of women and male academics is attributed to individual differences rather than to consequences of a male ordered world" (p. 134). A good example of the climate described by Bensimon and Marshall can be observed at the annual meeting of the American Educational Research Association where it was found that men spoke significantly longer than women and made significantly more responses to comments and questions

(Wiest et al., 2006). In addition, women's participation was lower in less structured aspects of the meeting than in aspects with greater structure. If these practices are attributed to individual differences rather than used as a catalyst to consider the structural ways in which we can encourage all voices to be heard, then we will continue to miss important voices in various national venues. Moreover, the larger educational policy environment is not limited to one conference or a specific act—it is enacted in a series of discourses, policy meetings, and reports; it is a cumulative effect where voices of women could potentially make a difference.

In *The Science Question in Feminism*, Harding states,

> Feminist politics is not just a tolerable companion of feminist research but a necessary condition for generating less partial and perverse descriptions and explanations. In a socially stratified society, the objectivity of the results of research is increased by political activism by and on behalf of oppressed, exploited and dominated groups. Only through such struggles can we begin to see beneath the appearances created by an unjust social order to the reality of how this social order is in fact constructed and maintained. (as cited in Fine 2000, p. 117)

This chapter explores how the social order of higher education policy is resisted by women in order to encourage transformative educational change and how some seek to maintain the dominant social order. There is research on resistance as self-defeating (e.g., that which helps to recreate the oppression) (Foley, 1990; MacLeod, 1987; McLaren, 1993, 1994) and resistance that focuses on transformation (Friere, 1970/2002; Giroux, 2001; Sorlorzono & Delgado Bernal, 2001; McLaughlin & Tierney, 1993). For the purposes of this analysis, resistance will be defined as that which focuses on transformation, as it is consistent with the perspectives of participants who shared narratives of resistance.

In the last two chapters, I explored the content (what) and process (how) of communication between higher education leaders in a national conference on higher education and the public good. Through examination of the linguistic complexities of this national discourse, I found that some women, people of color, and people outside the academy (i.e., community partners) were silenced and/or their perspectives reframed or discounted. The fact that women and people of color are silenced and/or their perspectives rejected or altered is not new (also see Chase 2005; Gilligan 1982; 1987; 1988; Green & Trent 2005; Rowley 2000; Smith 2004; Stanley 2006;

Tannen 1993; 1994). It is important to reiterate that the word "silence" in this context is to mean that ideas are shared but not centered in the discussion and/or not included in final reporting structures or revised models. As Tannen (1993) points out, silencing a person is not necessarily connected to volubility; a person may talk a lot or use many words to describe a concept and still be silenced. In the case of these earlier findings, a participant may or may not have felt silenced in the moment, but when their concepts and ideas are not incorporated in revised concepts of higher education's relationship to society, then the concept or idea is stifled, shut down, or silenced. Or if someone disagrees with the idea and the group moves on without addressing agreement or disagreement (i.e., "sweeping it under the rug"), then again a person may feel silenced.

In the discussion of chapter 5, I focused briefly on issues of race, gender, and community-university partnership relations, encouraging higher education leaders to "deal directly" with issues of educational inequity. In what follows, I take a closer look at the voices of women during these policy conversations. Specifically, I explore the voices of women who resist dominant ideologies and share "advocacy" perspectives and whose perspectives were reframed, redefined, and/or silenced in the discourse. In this chapter, I focus the micro-analysis more intentionally on feminist perspectives, answering my own call in chapter 3 for depth of analysis through the use of one theoretical perspective after the use of many.

I briefly describe the feminist lenses utilized for this analysis and offer three examples of women's voices of resistance to the dominant discourse of the national conference series. I situate the three examples next to each other in order to explore the complexities of gender discourse in detail. I consider both the content (what is said) and the process (how it is said) within this national policy context in order to apply women's change strategies that may have been missed in the larger discussion and consider the implications of these strategies on interrupting current patterns of educational inequity.

A Feminist Paradigm

bell hook's definition of feminism is one that has transcended generations of theorists. She states, "simply put, feminism is a movement to end sexism, sexist exploitation, and oppression" (2000, p. 1; also see hooks 1984/2000). This chapter takes a feminist perspective, drawing from feminist epistemology, theory, and methodology. Feminist epistemology—the philosophical grounding for deciding what kinds of knowledge are possible—addresses the connections between knowledge and its social uses and how patriarchal values have shaped the content and structure of that knowledge. Feminist theory is founded on three main principles (Ropers-Huilman, 2002). First, women have something valuable to contribute to every aspect of the world. Second, as an oppressed group, women have been unable to achieve their potential, receive rewards, or gain full participation in society. Third, feminist research should do more than critique but should work toward social transformation. Further, taking a feminist methodological perspective requires "examining how knowledge and power are connected, and so making visible both the hidden power relations of knowledge production and the underpinnings of gender" (Ramazanoğlu, 2002, p. 63).

Following a feminist paradigm, this chapter builds on previous chapters to address (1) the content of women's discourse about higher education policy, (2) the structure or process of these discussions, and (3) educational change strategies based on women's perspectives.

It is also important to note that there is a diversity of feminist thinking. In her book *Feminist Thought*, Rosemary Tong (2009) describes the conceptual roots of liberal feminism, parcels out the various nuances between radical-libertarian and radical-cultural feminism, and reflects upon the classical and contemporary Marxist and Socialist feminist perspectives. In addition, she describes psychoanalytic, care-focused, multicultural, global, postcolonial, eco, postmodern, and third-wave feminism (also see Nicholson & Pasque, in press). To be sure, the complexities of feminist thought cannot be reflected in a t-shirt or bumper sticker, but Tong offers a thick and rich debate between perspectives that work well together and those that contradict each other.

For the purposes of this facet of the study, I pull from feminist perspectives that focus on both the macrocosm of patriarchy or capitalism and perspectives that focus on the microcosm of the individual (Tong, 2009). In this way, I am able to focus on the specific discourse as found in the national policy conversations (micro-analysis) and connect the implications to the larger policies and practices of today's diverse society (macro-analysis). Deborah Tannen is an example of a researcher who pulls on macrocosm and microcosm feminist perspectives simultaneously. Tannen's (1993) research focuses on, for example, how women and men communicate their identity through the use of various linguistic strategies. She explores language and topics such as solidarity, indirectness, interruption, silence, power, and conflict. In this way, she explores performances of gender identity in specific contexts and yet connects these practices to the larger contexts.

In addition, I focus on feminist methodological perspectives through the use of narrative analysis. Narratives serve to highlight culturally recognizable explanations or interpretations, attend to the knowledge and intent of listeners and the protagonists in their stories, make use of a culturally commonsense epistemology, and take a moral or evaluative stance relative to the events in the story (Bruner, 1990; also see Walton & Brewer, 2001). In addition, people perceive and enact a construction of self based upon their belonging to specific social contexts (Monaci, Magatti, & Caselli, 2003). More specifically, I explore the voices of women in a situated discursive practice (policy discussions) that are tied to context-related enactments of identity and society (self and higher education).

Moreover, narratives frame our understanding of an identity of the self or of something or someone other than ourselves (Walton, Weatherall, & Jackson, 2002). The cultural practice of storytelling about personal experience constitutes a significant means through which people position themselves in relation to others and are positioned by others (Harre & Slocum, 2003). Narrativity also constructs and is constructed through practical reasoning. In this study, the participants' practical reasoning during the national policy conference series conceptualized the connections between higher education and society and models for change. Stated more simply, the women shared their perspectives on how to make strategic change in education. I illustrate how everyday, practical reasoning (what is said) produces various constructs for change (how to make change) and is produced by social interactions (who gets listened to), in which gender plays an important role.

Building off earlier chapters that have found: (1) women and people of color are often silenced or perspectives are reframed and (2) people with alternative perspectives (across social identities) often have their perspectives marginalized in the national policy discourse; this chapter explores the voices of women who fit both categories. In particular, I explore the discourse of three women who spoke during identified "peak" sessions across all four three-day conference series and shared perspectives that resisted the dominant paradigms of the conference series. These are representative examples of discourse by women in this series, and in-depth exploration of a few cases allows for more profundity in this analysis of the policy discourse and expands upon the pilot study (Pasque & Rex, 2010) and examples shared in earlier chapters, yet from an intentional feminist perspective.

FINDINGS

A number of women resisted the dominant discourse at different points during the entire national policy conference series on higher education and society. In each case, the content of what the woman offered was different from what was presented earlier by organizers. Each time a woman offered an alternative perspective, the content of what women said was reframed, redefined, or silenced by participants, organizers, or in final reports. Stated another way, alternative perspectives that attempt to address educational inequity and social injustice are neither "heard" nor centered in policy discussions or final reporting structures. In addition, the process by which the women who resisted the dominant narratives spoke up was different across the various women who spoke. For example, one woman used a "bridging move" that connected academic and non-academic language, another utilized more of a one-down approach to challenging the dominant models (as opposed to a one-up approach), and the third spoke with passion and emotion in order to ensure her perspective was heard.

In this section, I share three representative examples of the various discourses from women who resisted the dominant perspectives and explore language (1) of the women, (2) between women and men, (3) between women and women. I consider each of the challenges made by Judith, Stephanie, and Courtney through the theoretical and methodological lenses used in this study. I offer a reflective analysis that speaks to the women's potential strategies for educational change as shared through the national policy discourse.

Redefining Judith

This example explores the dialogue that ensued after Judith's plenary speech on the topic of how to make change in the relationship between higher education and society. Judith is a white woman director of a national policy organization on education and a professor. Relevant to this discussion is that conveners used the familiar ecological or systems model illustrated in figure 5.1, chapter 5. The organizers described the focus of this national conference series as the relationships between higher education and society, or the crescent between the "system of higher education" and "society" levels. The ecological model and other models were presented in the material that participants received prior to their arrival, mounted on a poster placed on a tripod in the room, mentioned by organizers in each conference opening session, and referred to by organizers and participants throughout the conference series.

Judith sparked the discussion by resisting this model and calling for differentiated change. Specifically, she disagreed with the conveners about *where* change should be defined. Judith stated,

> And I think the "just doing it" has to be—I guess it's the first thing, very differentiated. Because if you're a university president, you're going to "Just do it" [Nike reference was referred to earlier in the discussion] in a very different way than if you're a tenured faculty member or if you're a graduate student or if you're in some other kind of role. And for that reason, I think that in a bizarre way, that we have to violate one of the principles that's set out up here on the chart [see figure 5.1] and that's about "Systems Perspective." My suspicion is, and my experience is, that this [change in higher education and society] can't and shouldn't be systematic—that there's an inherent contradiction—that it's got to be idiosyncratic, opportunistic, differentiated. And, that maybe in the end, that's a different way of thinking systemically, but it liberates us from the, "Let's sit down and get a vision and a strategic plan and figure out what next steps are and what are our benchmarks," and all those things which I'm convinced will never get us to working for the public good in a way that's very different from the status quo.

In this narrative, Judith identifies the role differential between college presidents, faculty, and graduate students. This statement situates Judith in more of an advocacy framework where power is recognized and different strategies for making equitable change in policy on the topic of higher education and society are employed.

From here, the majority of speakers who followed positioned themselves either in agreement with Judith (in favor of differentiated methods of change) or in disagreement with Judith (change should be conducted in a systematic manner) or they viewed change as requiring both a systems and differentiated method. The effect was a contentious polarization of the two positions. In fact, eight of the eleven speakers who followed Judith in this session took up this topic to position themselves in relation to the issues she framed. The next four segments of the discussion demonstrate the views of people who positioned themselves in agreement or disagreement with Judith, or of people who reframed what she was "really saying" to include both the systems and differentiated perspectives (for more detailed analysis see Pasque & Rex, 2010).

> Nicole: [To Judith] I too feel the tension between the good yet idiosyncratic differentiated activities that are out there—and the sense of urgency I feel—and yet, also a sense of hopelessness about a systemic change.

In this example, Nicole, a white woman director of a national institute on community dialogues, aligned herself with Judith in support of the "good yet idiosyncratic differentiated activities." Nicole took this support one step further to state the sense of "urgency" and "hopelessness" about change that she felt when a systemic model is utilized. This statement mirrors other comments of support by women in this discussion. In the next example, Michael positions himself in disagreement with Judith.

> Michael: I'd come back to the "systemic part" just for a moment. I didn't agree with everything you said, [Judith], but I think in the conversation and particularly in [Kenneth's] comments, the central nub of this is that there are many different definitions of systems and then language that has evolved in the last decade or so as this becomes more public—and I'm a systems scientist so these are sensitive issues for somebody like me. [Laughter].

In this example, Michael, a white male professor, started his narrative by stating that he "didn't agree" with everything Judith had said. He offered that there were "different definitions of systems" and implied that her definition of systems might be limited. He couched this opposition with humor while naming himself as a "systems scientist," thereby aligning himself with the systems perspective and positioning himself (and not Judith) as an authority on systems. This is a subtle way in which power operates in national policy conversations, where Michael devalues Judith's comments while positioning himself as the authority.

Below are two examples of participants locating change around higher education for the public good as both a differentiated perspective (like Judith) and a systems perspective (like Michael). These two men illustrate ways to bridge the two perspectives.

> Angelo: I agree with everything you [Judith] said, the only part that I would perhaps want to talk about more is your mentioning about systemic maybe not being the way to go, and I hope, I think, what you're really saying is that maybe one has to work simultaneously from different perspectives. There is something to be said about the kind of organic "Just doing it" kind of approach.
>
> Kenneth: I think that the theory is to construct big boats with tall masts just over the horizon, and then to encourage them to sail into port by which time they've filled the field of vision of everybody who's standing on the land. And so I guess I don't see a distinction between idiosyncratic, entrepreneurial adventures and systemic change as long as you create the wind to make one lead to the other.

In the first example, Angelo, a male president of a national foundation, reframed what Judith was "really saying" in order to redefine the concept not as a bifurcation but as a combination of multiple perspectives, thereby creating a third definition. In this example, Angelo takes the liberty of redefining what Judith is "really saying" as though she may not have known herself or may not have been clear enough with her description. Kenneth, a white college president, reached a similar point by stating, "I don't see a distinction between idiosyncratic, entrepreneurial adventures and systemic change." His narrative also created a distinct and alternative repertoire in which differentiated and systems perspectives coexisted as boat and wind. More specifically, during the policy conference men advocated for the ecological/systems model for change—or both a differentiated and systems model—while women advocated purely for a differentiated model for change.

Judith's question about whether the ecological model sustains the status quo rather than supporting change is an important one. Tactically, Judith invoked a slogan from advertising that was used earlier in the conference as the conceptual framework for change but applied it in an external way consistent with academia. Judith assumed a different idea about making change and presented it in a discursive genre familiar to policy leaders in higher education. By articulating a minority position and casting it in language the majority could hear, her discourse appears politically strategic in this context. She has given voice to a position that might be heard but not extensively discussed in such settings. Other female participants, such as Nicole, took up Judith's strategy and continued to push against the dominant ecological model.

Further, Michael positioned himself as an authority of systems and simultaneously tried to reduce Judith's influence in this context. Michael may have chosen humor as a form of minimization—a way to accentuate the hierarchical positioning (DeVito 1992; Tannen 1994). In addition, Angelo redefined what Judith was "really saying." Angelo may have intended to help Judith during a volatile conversation by modifying her main point (Tannen 1993) or wield power in the situation through his redefinition (Tannen 1993). Kenneth, however, used a metaphor to disagree with Judith's differentiated model and find a compromise between the two approaches. The boat with the tall masts did not redefine what Judith said or reduce her authority but offered a differing perspective delivered with the dominant intellectual approach. The intersection of gender, power, and humor in this policy context is particularly noteworthy.

These four examples demonstrate the various ways in which higher education leaders located change in the relationships between higher education and society by constructing their experience and observations through a differentiated perspective, a systems perspective, or a combination of the two. Judith offered the differentiated model, which the women accepted and some men rejected or outright redefined. This redefinition or rejection of Judith's idiosyncratic—or alternative—method of change furthers Michael's perspective and limits the options of policymakers as they work toward transforming education. This taking of positions reflects the group dilemmas that play out in different "peak" sessions, as explored in the next example.

Reframing Stephanie

In an earlier chapter, I introduced Stephanie, a white woman professor of sociology, who also challenged the content of the organizers model. Her challenge of the model is useful in this analysis and may have been prompted by the challenge of Bob, a white male full professor of sociology. In the discussion, Bob states, "And I noticed the psychological [ecological] impact model does not have a team, a small group component. So, individuals, to me, are isolated. That's why they are individuals." With this comment, Bob explicitly shows how this idea of "team" is absent in one of the primary models utilized by the organizers of the conference. Stephanie returns to this idea approximately one-third of the way through the discussion and states,

> I think that as a relative newcomer to *this* level of meeting, I have been very excited about what feels like, although I'm not sure, an agreement that the organizing [ecological] model, the social movement model is, maybe, a best way to do this. And what, what would be useful to me, and maybe... I'm not thinking that that's always the model that I see that many institutions, I think, including my own, are necessarily using to develop this work. So, I feel that the [organizing organization] in this group could make a contribution if that was part of the shared vision and purpose. And what would be useful to me is if somebody could contrast what are some of the other models of change, which I sometimes hear, kind of, taken for granted. I'm thinking of entrepreneurial models, for example, more, forgive me, top down models for change. Whereas I have tended to agree that the way you brought about change was from the grassroots and organizing, but I don't think that's been taken for granted in the civic engagement [community].

Stephanie starts this comment by putting herself in a one-down position, stating that she is a newcomer to this level of a national meeting. She goes on to share her excitement about an "agreement" about the organizing model for this conference series. In this move, she articulates how organizers and participants have, up until this session, agreed upon the model. Stephanie mentions that this model may be "a" or one way to think about the topic at hand. The word "a," which she verbally emphasizes, is a key distinction. It signals to listeners that she believes this is one of *many* models that could be used.

Stephanie hedges (uses a verbal pause or repeats words, which is different from a stutter), Stephanie states, "And what, what would be useful to me, and maybe..." This hedging reduces the strength of her request—that the organizers present more than one model for considering this type of work. Stephanie resists the model presented during the conference with her request for additional models. In this indirect way, Stephanie resists the dominant perspective presented by the organizers—that this ecological model is *the* way to make change in the relationship between higher education and society.

In a related example, a woman of color graduate student also resists the dominant discourse in the conference, but prior to stating her opposition, she starts by hedging and putting herself in a one-down position by saying, "Um, I just wanted to say a couple— we've been talking about social movements and some of these side movements—I mean, I still have a lot to learn, but..." Further, a women who supports the dominant discourse begins her disagreement by stating that her position sounds "Pollyanna-ish." These non-direct manners of engaging in conflict are consistent with the work on gender discourse by Gilligan (1982; 1987; 1988); many of the women engaged in this national policy discourse used non-direct manners of engaging conflict. Gilligan's "different voices" framework is particularly useful for interpreting face-to-face verbal conflict across gender (See Sheldon, 1993). Gilligan found that girls/women operate from a care orientation and try to seek agreement in the way they frame conflict resolution. Specifically, Stephanie frames her resistance to the organizer's model in a way that provides the organizers with a resolution—the option of presenting additional models for consideration.

Immediately following Stephanie's request for other models, the facilitator, a white male, and a white male tenured faculty member at a research institution have the following exchange. It starts by the facilitator calling on the professor, the same full professor who originally mentioned that the concept of "team" is not included in the ecological model:

> **Facilitator:** Bob.
> **Bob:** ____. I want to speak to that. There may be someone else who wants to speak to you're—
> **Facilitator:** What topic is this? Is this an important?

In this performance move, the facilitator ignores Stephanie's request and calls on the next person to speak. It is Bob, a man of power, who calls attention to Stephanie's request and asks that someone else speak to this request, at which point the facilitator not only asks what the topic is but also questions the importance of Stephanie's request. This exchange adds to the vertical complexity of the topic. More specifically, Stephanie tries to request additional models and resist the notion that this ecological model is the only model to use in this conversation. The facilitator does not hear or tries to ignore or skip over this request. In this performance move, Bob views the direction of the conversation as negotiable and steps in to make certain Stephanie's comment receives a response. Intentionally, this professor uses his power in the room in order to ensure that an organizer responds to Stephanie's request for additional models. Bob could be using his situational power as a male ally in order to make sure Stephanie's important point is addressed by the group (for more on allies see Reason, Broido, Davis, & Evans 2005). And/or, he could view Stephanie's request as one that supports his own resistance to the ecological model (the topic at hand) and help to make certain that the request is entertained as it furthers his own perspective.

The response from one of the white male organizers that immediately follows the above comments also adds vertical complexity to the topic of models of change.

> Joseph: We—we intentional[ly] built into everything you've read before you came here, that perspective. We think that this is the problem of a sort, or a set of problems or a nest of problems or a complexity that requires an organizing movement or perspective. I suspect your question about what other models are available, is not invitational but rhetorical, you didn't want us to elaborate those at this moment, I hope not.

Joseph's response to this request starts with "we." He could be using the word "we" to mean all of the organizers (thirteen people on the curriculum design team), the ten people who did the organizing and planning of the event (seven graduate students, one professional staff member, and two faculty directors who worked with this organization at the time of the series), the foundation sponsoring the event, and/or the two faculty directors of the organization. In any case, the use of "we" adds strength to the organizer's argument, stating that he is not alone in his perspective or in his identification of the problem.

The comment "I suspect your question" lets Stephanie know that he is interpreting her question in a specific way. He then describes his interpretation of her comment as "not invitational but rhetorical." This defines her request as something he would like—a "rhetorical" comment versus an "invitational" comment.

In the exchange that immediately follows, Stephanie and Joseph start to talk at the same time, until Stephanie stops talking when he says the word "one."

> Stephanie: That would be another conference.
> Joseph: Yah, yah, at least another seminar.
> TALKING AT SAME TIME:
> Stephanie: Individually, that would be helpful, you might mention one. [She stops talking about when Joseph says "one"].
> Joseph: I'm almost too fired up to deliver one, but, ah, but yes. Yes. We believe that part of the excitement of this work could be found in placing values along side a new way of thinking and working and pushing those two together. That—that might help us to sustain interest and engagement because people would be learning a new way of working at the same time that they're focusing on values, which they hold deeply, so—if you wanted insight into some of what preceded this that was it.

Stephanie gives in to Joseph by saying, "that would be another conference," but then requests at least one other model for comparison. The organizer provides his rationale for not sharing additional oppositional models by stating that he is "almost too fired up to deliver one" but then realizes that Stephanie is requesting he share at least one additional model—that is, she is not withdrawing the request.

In response to this request, Joseph again starts with the use of the word "we" without defining who is included in his definition of "we." He shares that the model is one that couples values with "a new way of thinking and working," which sounds exciting but is not represented in the model.

In this performance move, Joseph reframes the question that Stephanie asked by stating, "if you wanted insight into some of what preceded this." Stephanie never requested more information about the preceding conversations that led to the use of this model. She requested different models of change. This example connects to additional research by Gilligan (1982; 1987; 1988) that found that boys/men have a need for an external structure of connection. In addition, boys/men tend to step back from the situation and appeal to reason, often losing site of the needs of others. In this example, Joseph connects himself with others by often using "we" in his response. This provides him with an external structure of connection—the group of organizers. In addition, he provides a rationale that is different from what Stephanie requests, thereby losing site of what she states she needs. He resists Stephanie's challenge of the models and uses his power as an organizer to reframe her request in a polite yet political manner. Further, the challenges to the organizers' models are not incorporated into revised models for change; the models remain unaltered at subsequent conference sessions and in conference summary reports.

Silencing Courtney

In the middle of a different session, Courtney, a Latina graduate student, does not challenge the model but instead challenges the dominant narrative in the session. This example is mentioned in a previous chapter through a different analytical lens and remains important to this analysis as well. Courtney challenges the dominant view, not that of the organizers, but of all the participants who have been engaged during the dialogue the past three days. She urges the group to name power structures *and* historic inequities, where italics indicate her original emphases.

> What sacred cows are we willing to slaughter? What pain are we willing to endure? I mean, people who participate in social movement think about risks all the time, they think about retribution. I mean, they pay with their lives, they pay with their futures, their reputations. And, I don't *really* hear us or see us talking about those things. We were talking about making change in very safe ways that allow us to maintain our status. That allows to maintain our privilege and our comfortability [Woman participant says, "that's right"]...we *have* to be true and we have to get up and admit we made mistakes. We have to get up and we have to tell the truth about things and we *have* to be willing to give some of our own power and our own privilege up in order to make things better for other people.
> Woman participant: Amen.

Courtney challenges the group's (organizers and participants) dominant ideology thus far by stating that participants are talking about making change in the relationship between higher education and society "in very safe ways." In her full narrative, Courtney describes truth telling as naming the historical and contemporary inequities that exist (e.g., the history of segregation) together with the structural ways to interrupt these inequities (e.g., diversifying the faculty). Courtney expresses that the group is not telling the truth in terms of naming historical and contemporary inequities and the structural ways to interrupt these inequities. With this comment, she advocates for continued efforts that increase sources of capital, as is highlighted by Bourdieu (1986), and for naming the power structures that sustain cyclical oppression, as is discussed by Foucault (1976).

Courtney goes on to emphatically state that social movements, such as the movement to strengthen the relationships between higher education and society, are about "fundamentally challenging the status quo"—something she stresses the participants have not done during this conference series. She articulates that even though they are engaging in conversations, participants in the conference are not addressing what they are willing to give up in order to make change. Participants are "talking about making change in very safe ways that allow us to maintain our status." Courtney resists the status quo by posing a direct question to participants through her use of the metaphor of the holy cow: "What sacred cows are we willing to slaughter?"

As Courtney becomes more assertive in her challenges to participants, she hedges less often and becomes more emphatic with her weight on particular words (as noted with the italics). Potentially, she raises her voice throughout the narratives to make certain her voice is heard. Her statement is met with verbal supportive gestures from two other woman participants of "that's right" during her statement and an "Amen" at the end of her statement. Courtney's performance move is a point in the session where the conversation topic changes (a horizontal performance move) and participants begin to talk more deeply about the way in which this group of participants is engaging with each other and with the topic of educational in/equity (vertical horizontal move), but only after a male speaker expands upon her comments. This is different from previous conversations, which focus on higher education's relationships with society in general or specific best-practices. However, even though Courtney's comments shift the conversation during the conference, her comments were not incorporated into the final reporting documents. This is one example of how alternative perspectives that challenge the status quo and encourage giving up power and privilege are relegated to the margins of the discussion. This finding is related to existing literature (Wackwitz & Rakow, 2004). Courtney's alternative frame is not included in follow-up sessions by the participants and organizers with dominant processing models, even though she received verbal support from some other women participants. In the final reports, some women's perspectives are silenced by omission.

Giroux (2001) argues that "resistance must have a revealing function, one that contains a critique of domination and provides theoretical opportunities for self-reflection and for struggle in the interest of self-emancipation and social emancipation" (p. 109). It is Courtney, a person who holds target identities (e.g., woman of color graduate student), who urges the higher education leaders with power to engage in self-reflection, take risks, and make change to the status quo. Further, to include the voice of women of color with that of white women in this sense adds an important antiracist feminist perspective that argues against assumptions made about gender by and about predominantly white women (MacDonald 2002); this perspective stresses the interconnections and complexities between gender, race, and class.

DISCUSSION

Parsons and Ward (2001) state that "academics are disciplined not to recognize the subtle, unconscious sexism that permeates the academy" (pp. 56–57). These three examples of individual women who resist the dominant models and narratives in the national academic policy discussions may appear unconnected and yet represent a pattern of institutionalized sexism displayed in a policy context even with participants who may have had the best of intention to embark upon transformative educational change (an assumption based on the stated goals of the series). Such resistance by women is found to be a political strategy toward change (Gilligan, Rogers, & Tolman, 1991) and provides an impetus to explore the complexities of gender in educational policy and alternatives to current paradigms.

Process and Content

In terms of *process*, Judith's strategy to "just do it" in order to strengthen the relationships between higher education and society is supported by women and redefined by men. This idiosyncratic strategy for change is shared in a familiar sound bite (Nike slogan) that connects academic and non-academic people in the room. In the second example, Stephanie requests alternatives to the foundational model and an organizer reframes her question to ignore the request and further the ecological model of the conference. In addition, her request is addressed only when a white male full professor steps into the conversation and asks that someone respond to Stephanie's request for additional models. This male serves as an important ally in order to stop the facilitator from moving past Stephanie's request. Further, the fact that she persists with her request for additional models (requesting at least *one*) ensures that her request is answered, even if not directly. In the third example, it is when Courtney becomes more emphatic and potentially "angry" when questioning the ways in which participants are maintaining their privilege and status in this national policy discussion that woman participants verbally support her. Her challenge to the group is silenced at the end of the conference by participants not following up on her comments and not including them in final reporting documents.

As Marshall (1999) notes when discussing educational policy,

> people use speech as a power tool to create power, to effect a desire or goal, and to block, resist, and create opposing strategies (Ball, 1990; Foucault, 1981). Privileged speakers' truths (and policy analyses) prevail; a "discourse of derision" can be used to displace or debunk alternative truths (Ball, 1990). Research on how marginal issues get into the public discourse is about gender and about democracy. (p. 65)

In this case, both women and men use speech as a power tool to communicate their perspectives. However, when women challenge dominant ideologies, the discourse of derision specifically redefines what women are "really saying," reframes questions that challenge dominant models, and silences perspectives in order to maintain the current social order. As mentioned, silencing is not necessarily connected to volubility (Tannen 1993); omitting comments from the final reporting documents is one method of silencing perspectives in a policy context.

In addition, the narratives from all three women include examples of hedging and two include examples of one-down positioning. These nonverbal pauses or non-direct manners of engaging in conflict are consistent with the findings on gender discourse by Gilligan (1982; 1987; 1988). Coupled with these hedging moves are the women's own attempt at leadership and educational change. In a related example, higher education researchers Astin and Leland (1991) offer a feminist conceptual model of leadership that "rests on the assumption that leadership manifests itself when there is an action to bring about change in an organization, an institution, or the social system—in other words, an action to make a positive difference in people's lives. Leadership, then is conceived as a creative process that results in change" (116). Each of the women attempts to bring about change in this policy conversation or in educational policy as a whole. As such, Judith, Stephanie, and Courtney would be included within Astin and Leland's definition of feminist leadership. From here, the educational policy questions are thus redefined: What is the content of what these women share even as their comments are being reframed, redefined, or silenced in the national discourse? What is the content or alternative perspective that each woman offers for consideration?

In terms of *content*, Judith questions the foundational ecological model and talks about differentiated change to strengthen the relationships between higher education and society. She describes the "just do it" strategy as having multiple and simultaneous entry points. Her idiosyncratic perspective is much like Al Gore's approach toward addressing global warming. In myriad speeches and the movie *An Inconvenient Truth* (David, Bender, & Burns, 2006), Gore discusses the layered complexities of global warming and then offers numerous strategies to simultaneously address this problem with a worldwide approach. These multilayered global efforts require leaders from various industries and countries to come together to address these critical issues. Gore's efforts have led to a Nobel Peace Prize in 2007. The ripple effect of increased awareness about global warming can be witnessed daily, such as with the demand for energy-efficient cars, recycling bins throughout campuses, and pervasive "green" marketing strategies on water bottles, t-shirts, and cloth bags, or on children's television shows that introduce the topic of reduce, reuse, and recycle.

In a similar vein, Judith talks about differentiated educational change as an issue for all, where one collective strategic plan will not necessarily interrupt the status quo. Such an idiosyncratic strategy requires many higher education leaders to "just do it" from multiple approaches, with multiple constituencies in mind. In this way, Judith offers a new way for political strategists and education leaders to consider the layered complexities of educational inequity and provide myriad strategies for change.

This idiosyncratic approach toward change also reflects the New Science perspective. As Wheatley (1999) describes,

> In organizations, which is the more important influence on behavior—the system or the individual? The quantum world answered the question for me with a resounding "Both." There are no either/ors. There is no need to decide between two things, pretending they are separate. What is critical is the relationship created between two or more elements. (pp. 35–36)

Expanding on the New Science perspective, Love and Estanek's (2004) discuss its relevance to student affairs in higher education. They argue for a "both/and" perspective where the Newtonian science and the New Science of chaos and unpredictability are combined in dialectical thinking. Their conceptual framework for the future of organizational behavior in student affairs also includes transcending paradigms, recognizing connectedness, and embracing paradox. Again, this idiosyncratic approach focuses on multiple strategies enacted in interconnected contexts that create a ripple effect of change: a potential strategy to strengthen the relationships between higher education and society.

Stephanie's contention also questions the foundational ecological model for change in the relationship between higher education and society. She requests the group to consider additional models from a more of a "grassroots" organizing perspective. It is important to note that Stephanie did not attend the earlier conference series where Judith provided an alternative model. The organizer (and a few other people in the room) did attend both sessions and chose not to resurrect Judith's differentiated change model. Instead, the organizer reframes Stephanie's question to provide "insight into some of what preceded this" conference series: a verbally strategic tactic to alter her question and then answer a different question. This is a tactic utilized often by politicians, business leaders, and sometimes our close friends when asked a question they do not wish to answer. Further, Stephanie's request is addressed only when a white male ally speaks up and suggests someone answer her question. The vital role of allies in educational policy contexts cannot be underscored enough. It was only with the support of this ally that Stephanie's request was acknowledged as the facilitator did not pick up on her request for additional models.

Stephanie's request for additional grassroots models suggests she disagrees with the ecological organizing model and is looking for a contemporary and bottom-up strategic model for change. Such a model reflects humanitarian and activist efforts sponsored by organizations and community groups from around the world. For example, whether you sported a "Team Jolie" or "Team Aniston" t-shirt in 2005, one cannot deny the humanitarian efforts of Angelina Jolie. As a UNHRC goodwill ambassador (the UN Refugee Agency), Ms. Jolie presented a plea to CNN viewers of Anderson Cooper's 360 to help the 2 million refugees in Pakistan, where the number of refugees has been growing daily (UNHRC, 2009). This has been reported as the largest human displacement in a decade, since the devastation in Rwanda, and Jolie is adding her voice to help raise awareness of this issue.

Most UNHCR operations are in the field, managed from a series of regional offices, branch offices, suboffices, and field offices. Although the worldwide operation has become highly complex, ranging from recruitment of new staff and ensuring their security in dangerous situations to the procurement of everything from medical supplies and bulk food shipments to aircraft charters, the organization still relies on the efforts of hundreds of people working in the field directly with people in local communities all over the world. These people include famous ambassadors and the many viewers of CNN who may be encouraged to work with efforts to aid in the refugee crisis. A global organization that may not fit with Stephanie's definition of "grassroots," the UNHRC, however, is a model of how a large complex organization can choose knowledge and action over gridlock (such as the tragic inaction immediately following hurricane Katrina). In this way, this model connects "grasstops" and "grassroots" efforts for change.

Grassroots organizers are not limited to actresses and those they entice to get involved in local communities. Author Alice Walker is known for the awareness she brought to the issue of female mutilation through her book and movie of the same name with Prathibha Parmar, *Warrior Marks* (Walker & Parmar, 1993a; 1993b), as well as her ongoing community efforts on environmental and economic justice. Journalist Lisa Ling has also brought awareness to numerous crises such as bride burning in India, gang rape in the Democratic Republic of the Congo, child trafficking in Ghana, the aftermath of hurricane Katrina, and her sister's captivity in North Korea. Humanitarian and activist efforts such as these are included in the global feminist perspective where global feminists "challenge women in developed nations to acknowledge that many of their privileges are bought at the expense of the well-being of women in developing nations" (Tong, 2009, p. 8). Such models for change focus on individual human stories together with the larger social benefits, all within various sociopolitical, historical, and cultural contexts.

Finally, the content of Courtney's narrative asks participants to spend time engaged in "truth telling" and to name strategies for change that may not be as safe or allow participants to maintain their current status. Making visible one's own power and privilege is a difficult task, particularly for people in privileged positions with agent identities (Wildman & Davis, 2000). In the work of Adams, Bell, and Griffin (1997), this is the first step toward engaging in difficult intergroup dialogues about gender, race, class and individual, institutional and systemic oppression. Such a step prepares one for interrupting the cycle of oppression and embarking on systemic change (Harro, 2000). Courtney's strategy follows this model and may help to strengthen trust. It starts with identifying privilege and admitting the mistakes that have transpired in the educational system in the past. This process of identifying the current system of inequity becomes the starting point for interrupting the dominant paradigm and crafting an action strategy for change.

While Stephanie's model for change may focus on principles from a global feminist perspective, Courtney's strategy focuses on a multicultural feminist strategy. Similar to global feminists, multicultural feminists focus on women's varying social, cultural, economic, and political contexts. Multicultural feminists, however, "focus on the differences that exist among women who live within the boundaries of one nation-state or geographical area" (Tong, 2009, p. 8). In Courtney's case, her objectives shed light on the historical current inequities in US education—the ways in which policy leaders perpetuate the status quo—and encourage leaders to break the cycle of oppression.

Dilemmas for Policy Change

Each form of resistance poses a dilemma for the group as a whole when the participants and organizers do not agree. In each case, the content of the dilemmas are not captured in recrafted models or in revised visions for change that hope to strengthen the relationships between higher education and society. Instead, the dilemma is couched in a process of "broad, if not universal, agreement" in the final reporting mechanisms for the conference series. The final report serves as a strategy to show consensus, whether present or not. The ecological model starts off as and remains the focus of the ecological model during this educational policy series. With this rhetorical move in the final report, organizers continue to ignore the alternative concepts of differentiated change and requests for additional grassroots models. In this manner, the dominant cognitive processing models continue to be perpetuated during the conference, throughout organizational processes, and with the formal documentation of the conference series.

Dominant discourse can subdue, redefine, reframe, and resist divergent perspectives and represent a "consciousness" of general support. In this sense, organizers universalize discourse to coordinate people's diverse perspectives into a unified frame (also see Smith, 2004). This is not to say that all alternative perspectives are "better than" the dominant perspectives; however, consideration of multiple options opens up viable opportunities from which to interrupt the current cycle of educational inequities. As is the iterative process with policy discussions, this is not where the organizers left the conversation; they have continued to work through other venues to try to strengthen the relationships between higher education and society, with each other and independently. Some of this work continues to utilize this ecological model and other work introduces new models for change.

As Bensimon and Marshall (2000) note, feminist perspectives of higher education policy demand new agenda-framing and, as this chapter shows, also require facilitation that supports and not redefines or marginalizes feminist and alternative perspectives. In this manner, an emergent and inclusive process within policy discussions may support new content and an organic form of democratic principles, which may provide alternative action strategies for educational change. Further exploration of *Reconstructing Policy in Higher Education* from a post-structuralist feminist perspective can be found in the new edited book by Allan, Iverson, and Ropers-Huilman (2009).

In a performance ethnography of the narratives shared in this chapter, I equate the three narratives of resistance to the story of the *Three Little Pigs* (Pasque, 2009). To the surprise of some in the audience, it was not the organizers who were represented as the wolf or the three women who served as the little pigs. In this version, it was the organizing team who constructed the houses (models) and it was the women who "huffed, and puffed, and blew the house down" or at least attempted to blow the house down. This rendition of the findings from Judith, Stephanie, and Courtney provides three additional points for analysis and reflection.

The first point raises a question about the implications of being taught fairy tales based on a dualistic "good" versus "evil" scenario. The wolf in the *Three Little Pigs*, the giant in *Jack and the Beanstalk*, and the wolf in *Goldilocks and the Three Bears* all teach young people about a dualistic paradigm of good or evil; right or wrong. The "Team Jolie" and "Team Aniston" example above follows this same principle; we must choose one side and demonize the "other." The ramifications of this paradigm are that we try to construct houses (e.g., models, a conference series, perspectives) that cannot be blown down. We reject any forms of resistance to our ideas as though alternative perspectives force us to choose. The complexities of the issues and shades of gray are absent. We end up demonizing the person who attempts to resist our ideas, or the original idea itself, rather than perceiving it as an opportunity to consider multiple perspectives. We often hold tight to our worldviews rather than considering a reshaping of the original form. Including the perspectives of people different from us enables us to see an issue from a standpoint of which we were not originally privy; we consider the perspectives of the wolf, the pig, *and* people peripheral to the story (i.e., crafters of musical instruments, personnel in the local building supply store, and harvesters of oats for porridge). This opportunity to view a different perspective increases the information we draw from as we strategize for change.

A second point that can be gleaned from the comparison with the *Three Little Pigs* asks, "What are the ramifications of building house after house with brick after brick?" The implications of this approach can be found in suburbia across the United States, where many of the houses are now sadly in foreclosure. The diversity of people and construction starts to diminish as people construct their neighborhoods. Houses become bigger and stronger in order to keep away the evil wolf, and gates are built at the end of the subdivision. People become fearful that their house will be blown down. In such a case, people become so concerned with constructing strong, airtight homes and wrought-iron gates (models) that they forget to engage with the community around them. Who do the gates keep in? Who do they keep out? What perspectives are included or excluded with such structures in place?

The final point that I will make is the connection between fantasy and reality. As a child, we explore the world of fantasy through books, movies, and playing with friends. Some of us continue exploring the world of fantasy through books (*Lord of the Rings* by J. R. R. Tolkien, 1965), movies (*Harry Potter*, book series by J. K. Rowlings), and virtual communities such as found in Wii and Second Life. How do we define these fantasy worlds and are they really exclusive of reality? If reality is socially constructed, then can people construct reality from an aspect of fantasy? Further, if you continue to operate from your own reality and not include the realities of others, does that not create an element of fantasy in your own world? It follows that if we construct our own reality, we also construct our own interpretations of fantasy and of things we do not experience in our own immediate realities. More specifically, I question whether this element of fantasy, or ability to "fill in" a gap within our own direct experience, helps perpetuate notions of the "other" that objectify people we do not know well. There is a way in which people sexualize or demonize people who are not within our immediate circle, be it people of a different race, ethnicity, class, gender, sexual orientation, ability, and/or age than ourselves. Does that interpretation—which may be based on fantasy—become a part of our own reality and, in turn, impact the ways in which we act upon and within the world? What are the implications of the connection between reality and fantasy for the policies and procedures crafted, for people like us and not like us, in the field of higher education?

CONCLUSION

Acknowledgment and inclusion of various perspectives of the relationships between higher education and society is paramount as leaders make more informed choices about how to work toward systemic and equitable change in education. Understanding multiple models, the ecological model and revised models based on alternative perspectives presented, provides multiple frames for considering the relationships between higher education and society. This inclusive information becomes important as leaders strengthen arguments for effective policy change, particularly for developing change strategies that address educational inequities across gender, race, and class. If organizers and/or participants dismiss alternative perspectives offered by participants through discourse processes and pay attention only to the dominant frames of understanding, then they will not have an inclusive understanding of the problem to the exclusion of particular educational change strategies.

As Wackwitz and Rakow (2004) note, women are too often

> denied access to communicative forums—interpersonal, group, organizational, and mediated—or admitted to them only to have their ideas dismissed out of hand as deviant or irrelevant. To have voice is to possess both the opportunity to speak and the respect to be heard. (p. 9)

This research supports the findings of Wackwitz and Rakow and expands our knowledge about the complexities of gender in national education policy discussions. The findings encourage a revision of communication processes and an exploration of content offered by women so women may be "heard."

CHAPTER 7

A Tricuspid Model of Advocacy and Educational Change

Susan Komives (2000), the current president for the Council for the Advancement of Standards in Higher Education (CAS), put out a call to awareness for higher education leaders "to be conscious of our own gaps" (p. 32) between knowledge and action. She states that, "[educators'] systematic processes too often stop at the acquisition of knowledge. The much harder and more meaningful process is to facilitate understanding and wisdom, leading to the intentional self-authorship inherent in informed thought and action" (p. 31).

Higher education leaders need to "inhabit the gap" between knowledge and action by taking what we currently know about educational inequities and transforming this knowledge into everyday decisions as we set new direction for the future of higher education and the public good. But, if we have yet to understand higher education and community leaders' myriad perspectives about educational inequities and continue to marginalize perspectives in national policy conversations, how may we further our own knowledge that leads to such action?

In this chapter, I address this connection between knowledge and action. Initially, I provide a brief overview of the study's findings regarding the discourse of leaders. Based on this knowledge, I offer the Tricuspid Model for Advocacy and Educational Change, an emergent action model that focuses on organizational behavior and discourse. If such a model was employed, then concrete ideas in terms of *how* to make change in higher education may not have been overlooked in the discourse. Action strategies such as Judith's idiosyncratic method for change, Stephanie's grassroots models, and Courtney's truth telling examples to develop trust (see chapter 6) would have been centered and provided additional options to connect knowledge and action. This emergent action model is useful for leaders within higher education and may also be instructive for leaders in various professions across the globe.

Finally, in this chapter I focus on steps for future research. This section outlines the boundaries of this current study and provides options for researchers to consider as outgrowths of this research.

Overview of Findings

To reiterate, when exploring the complexities of this question about higher education leaders' perspectives, it is important to bridge the macro-analysis of the literature with a micro-analysis of face-to-face discourse. As Miller and Fox (2004) state,

> These [macro- and micro- bridging studies] remind us of the complexities of everyday life and how agency and constraint are simultaneously implicated in it. An exclusive focus on either side of this dichotomy is inadequate, since everyday life is lived within culturally standardized discourses and the discourses are change by the ways in which we use them. (p. 47)

Four cognitive processing models, or frames, were found through the macro-analysis where authors and speakers describe conceptualizations of higher education for the public good. The four frames include the *Private Good, Public Good, Public and Private Goods as Balanced*, and *Public and Private Goods as Interconnected /Advocacy.* To briefly recap, the *Private Good* argument addresses issues of economic injustice but does not connect this injustice and action with historic and contemporary cyclical oppression. A *Public Good* argument concentrates on higher education's contributions to public and social goods such as adding to a diverse democracy, yet it is void of economic discussion and often ignores the inherent power of the market. The *Balanced* perspective discusses both, and yet the public and private domains remain mutually exclusive. It is the *Interconnected/ Advocacy* perspective—where the public and private good are understood as interconnected—that addresses cyclical oppression around race, class, and gender and works toward interrupting the current paradigms of educational inequities.

In the micro-analysis with leaders behind closed doors, two themes emerged: (1) participants consistently challenge the organizers' foundational models and (2) participants challenge the dominant discourse of the conference. First, a few higher education leaders challenge the content of the models presented by conference organizers in earlier sessions, in handout material, or on large posters placed around the room. It was primarily women and people of color who stated such challenges; however, one white man also challenged the organizers models. Following such a challenge, other participants or organizers added to this challenge, rejected the idea, or did not take up the challenge in any way. These findings are supported by earlier research (Pasque & Rex, 2010). The challenges to the organizers' models are not incorporated into revised models for change; the models remain unaltered at subsequent conference sessions and in conference summary reports.

The second primary finding is that participants of color challenge the content of the dialogue engaged in by participants and organizers. In particular, a few of the participants of color argue that the large group of organizers and participants perpetuate the status quo around educational inequity with their discourse and are not talking about making substantial change for the public good. The content of language used by these participants echo the *Advocacy* frame as described in the macro-analysis. These challenges to the dominant narrative are also not taken up or are rejected by white participants or white organizers. There are a few points in the discourse when men of color from the academy make strategic bridging moves that include the *Interconnected/Advocacy* perspective into the discussion. These combinations of moves are tactical in that they support the person but do not necessarily agree with the content of what is said or alienate people with a dominant perspective. These moves are needed in order for marginalized perspectives to be considered in the discussion.

A more in-depth analysis from a feminist perspective explored both the content and process of the discussion. Participant perspectives by women with an advocacy perspective and dialogic processes that ensued were explored in detail. The content of the discourse that was missed by the marginalization of certain perspectives was expanded upon as ideas for change.

Each challenge to the discussion about how to strengthen the relationships between higher education and society poses a dilemma for the group as a whole; participants and organizers do not agree. In each case, the dilemmas are not captured in recrafted models or in revised visions for change that hope to strengthen the relationships between higher education and society. Instead, the dilemma is couched in "broad, if not universal, agreement" in the final reporting mechanisms for the conference. These alternative voices are present in the national discourse but are often relegated to the margins in higher education policy conversations and literature about how to strengthen the relationships between higher education and society. Here lies the crux of the "disagreement" between higher education leaders in the national discourse. The alternative cognitive processing models consistently come from people who clearly argue for a change in the perpetuation of the status quo and who address the content of the current dominant models, such as the *Private Good* frame.

Taken together, the findings lead to the development of an organizational change strategy that supports the concept of justice and inclusion, the Tricuspid Model of Advocacy and Educational Change. But, first, I must make explicit that all frames, including the *Private Good* frame, address issues of justice, albeit in different ways. Further, I have no interest in inverting the old logic of the academic hierarchy and excluding white men (Bensimon & Marshall, 2003). I do have an interest, however, in highlighting cultural exclusions and developing inclusive space for emancipatory change.

The Tricuspid Model of Advocacy and Educational Equity

The *Interconnecting and Advocacy* frame continues to be marginalized in higher education policy discussions and written reports, so it is no wonder that the *Private Good* and *Balanced* arguments that rationalize public funding for higher education prevail. The *Private Good* frame remains the official verbal and written discourse in educational policy, to the exclusion of other ways of thinking. The perpetuation of the current trajectory throughout the field of higher education and the continued marginalization of alternative frames will be detrimental to actualizing educational access and equity, as it is based on a win-lose framework as opposed to a win-win framework.

Moreover, Kezar (2004) argues that if legislators, policymakers, and the public are unclear about why higher education is important to society, then other public policy priorities may gain support at the expense of higher education. With the collective findings of the pilot study, literature review, and the discourse at a national conference, I extend Kezar's statement. If legislators, policymakers, higher education leaders, and the public are only are exposed to dominant cognitive processing models such as the *Private Good* model, then they will be limited in scope and will not be able to critically make social change that addresses the current and persisting inequities in the educational system—higher education for the good of an inclusive public. Burkan (1991) identifies this as "paradigm blindness" or the failure to perceive one's own paradigm as a frame of reference.

As outlined in chapter 2, concepts of justice are included in all perspectives about higher education for the public good presented in this text (*Private Good, Public Good, Balanced, Interconnected/Advocacy*), however, if the public and private goods are presented as *Interconnected* and the language is offered from an *Advocacy* frame, then the idea is not included in—or fully adopted into—the dominant narratives in the field or in face-to-face discussions. In addition, it is often the strategic bridging moves that are needed in order for these marginalized perspectives to be considered.

To address the *Private Good* perspective specifically, this frame takes the stand that economic disparities exist and that if people (undefined) are educated through postsecondary education, they will be able to be economically viable for themselves, their families, and the local and state community. This market perspective delivers a certain conception of justice and equity. To reiterate, this cognitive model mirrors the "pull yourself up by the bootstraps" perspective—where all people who have an education should be in a position to be successful and financially contribute to the public good through their own individual success. A major premise for some with the *Private Good* perspective focuses on issues of fear of other countries; garnering funds for the survival of an individual, institution, state, and/or country; and the marketization of higher education. It is crucial to note that the *Private Good* frame does not specify *how* all people in US society will be able to gain access to higher education, *who* has yet to gain this access, or even that inequities across race and gender exist in the current system in addition to class-based inequities. The damaging implications of such a perspective are outlined in Michael Apple's (2006) *Educating the "Right" Way: Markets, Standards, God, and Inequity* as he thoroughly considers the current era of conservative modernization.

I argue that solely a *Private Good* market system of higher education has the potential to further stratify the system of higher education in terms of race, gender, and class by ignoring historical and contemporary inequities. For example, the proposal by Brandl and Weber (1995) described earlier that provides state appropriations directly to students does not ensure that the funds would reach colleges and universities; the funds may be diverted to postsecondary organizations or for-profit corporations. This sole market mentality has not proven beneficial in the long run for Wall Street or Main Street, so why mirror it on College Avenue? In addition, student grants may not increase with the cost of inflation and have the potential to be cut over time if other public enterprises take priority over higher education.

As state appropriations continue to be reduced and tuition increased, the public could imply that college is for elites as it is so expensive. Even though access to community college may increase through President Barak Obama's American Graduation Initiative (2009), college still serves as a sorting mechanism where access to community college does not always equate to the advantages received from an Ivy League or land grant flagship institution as spelled out in Brint and Karabel's (1998) *The Diverted Dream.*

Furthermore, research findings show that students of color and students from lower socioeconomic statuses (SES) are less likely than their peers to receive information about financial aid and funding opportunities (Kezar, 2005; Southern Education Foundation, 1995). Kezar (2005) argues that the high price of college, even if it is subsidized, leaves a "financial aid gap" or "sticker shock" (p. 32), which deters students from attending college or from attending a more expensive institution. Gutmann (1999) contends that a high-tuition, full-scholarship policy for students of color and students of need would promise both equity and efficiency in terms of increasing access to higher education. Further, Gutmann states that this may come with its own set of obstacles such as students not feeling a sense of belonging. She argues that by keeping tuition low, institutions also keep the external barriers for admissions low. Low-income students feel "more like equal members of a low-tuition university than they do with expensive universities where they are among a minority of students who are fully subsidized" (p. 224).

In sum, although a *Private Good* perspective contains a conceptualization of social justice, I argue that this perspective does not encompass an inclusive concept of social justice. This perspective is limited to an "input-output" argument and continues to stratify people across historical and cultural inequities, thus adding to the *cumulative oppression* across race, gender, and class that Lesley Rex and I (2010) outline elsewhere. However, a private perspective does provide important considerations when it is interconnected with a public perspective as, together, they provide a complex approach to higher education and the public good.

Further, "New paradigms are created/discovered while paradigms they are to replace are still successful. The person who shifts paradigms is almost always an outsider and new idea(s) are almost always rejected initially" (Fried, 1994, p. 38). Multiple models for change that push beyond the familiar economic model are vital as we address the current changes in higher education and society. Fresh, new perspectives must be explored in order to expand awareness and possibility.

The questions then change: How do higher education leaders offer forums on this topic in such a way that diverse and multiple perspectives are encouraged, voiced, supported, and included? How do well-intended organizers and policymakers center alternative cognitive processing models, such as the *Interconnecting and Advocacy* frame, during national policy discussions? In addition, how do participants with alternative cognitive processing models ensure that their voices are heard in these types of settings? Is this language that threatens the status quo too extreme, which is why a more tempered approach, such as the *Balanced* cognitive processing model, is more palatable in policy circles? (Yes, I have heard this argument from colleagues many times.) Is there an "acceptable" way to disagree with a model—or alternative models—so it is centered for the group to consider? Who determines what is "acceptable"? As Jane Fried (1994) points out,

> Outsider status gives the ability to see the dominant paradigm more clearly. It also offers the opportunity to share perceptions with colleagues and lead the way in establishing the legitimacy of alternative paradigms as a basis for conceptualizing and conducting the work of higher education. (p. 31)

In the hopes of legitimizing alternative paradigms as leaders work toward addressing critical issues in higher education, I offer a Tricuspid Model of Advocacy and Educational Change that emerges directly from the findings and that requires rethinking the role of power brokers, centering marginalized perspectives, and including voices of all (target and agent) people in society (See figure 7.1). The Tricuspid

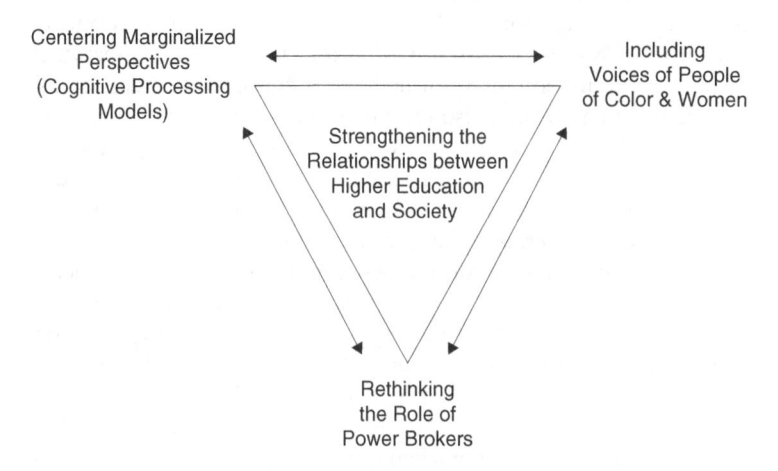

Figure 7.1   A Tricuspid Model of Advocacy and Educational Change

Model requires that all three areas and the connection between the areas are addressed in our national policy discourse. Although the model might appear provincial, as Komives (2000) aptly points out, it is often difficult for higher education leaders to "inhabit the gap" between knowledge and action.

Rethinking the Role of Power Brokers in Higher Education

Power brokers frequently come together to discuss policy and gate-keeping issues in the field of higher education. Whoever controls this organizing feature also determines, in part, who is invited, what issues are followed up on, and the final reports that are disseminated or, what I term, *information sifting*. As is shown through this study, it does matter who organizes such gatherings, what agenda and procedural models are followed, and what the cognitive processing models of the organizers and funders are. As was explored in the findings, it was the cognitive processing models (perspectives) of some of the organizers that often prevailed as the dominant model in the large group discussions on the national level. It was also the organizers who put together the list of invitees and speakers who held differing perspectives. In addition, an organizer, postdoctoral researcher, and invited speaker each played an important role of connecting dominant and alternative cognitive processing models by using important strategic bridging moves.

Higher education leaders need to create a climate where power brokers foster space for alternative visions around social justice and educational equity, not simply in the conference seminar rooms but also in the creation of revised models for change, final report documents, dissemination of information, and policy changes. Such a change in climate would also require a shift in the social balance and institutional discourse (Miller & Fox, 2004) so that people with alternative cognitive processing models, who directly address issues of power and privilege in the US system of higher education as related to issues of race, gender, and class, are centered.

It is also important to consider whether the network of practices in higher education needs the current problematic structures that perpetuate inequity in order to exist. As Fairclough (2001) mentions, questions need to be asked about "whether those who benefit most from the way social life is now organized have an interest in the problem not being resolved" (p. 236). Jones, Torres, and Arminio (2006) echo this line of questioning when they ask higher education researchers to "address not only what is being said but also what is not, not only what was said and quoted but also what is being protected from public view and why" (p. 31). This raises the question of whether national higher education leaders—we—benefit from the use of specific frames of higher education's relationship with society. More specifically, in what ways is social reproduction operationalized in this context? Does a market-driven frame that is often heard and accepted by policymakers and funders help support future funding possibilities? Are certain perspectives more palatable in final reports that are widely disseminated or in speeches made by people linked to specific national organizations?

On a related point, Bok (2003) contends that "the incentives of commercial competition do not always produce a beneficial outcome; they merely yield what the market wants" (p. 103). In US higher education, faculty experience a tremendous amount of pressure to acquire additional external funds. For example, Slaughter and Leslie (1997) state,

> to maintain and expand resources faculty had to compete increasingly for external dollars that were tied to market-related research, which was referred to variously as applied, commercial, strategic, and targeted research, whether these moneys were in the form of research grants and contracts, service contracts, partnerships with industry and government, technology transfer, or the recruitment of more and higher fee-paying students. (p. 8)

Current faculty and administrators are experiencing pressure to alter agendas in order to guarantee financial support and succumb to the market. In *Academic Capitalism and the New Economy*, Slaughter and Rhoades (2004) describe further the theory of academic capitalism and warn people in the field to be more cognizant of these influences. The absent presence of the current era of conservative modernization in education must continue to be highlighted, its ramifications illuminated, and alternatives proposed.

A dilemma also faces educational activists for equity relative to participation in such national conferences and steering committees. As it stands now, there is a consensus strategy that operates in the collection of opinions and ideas of participants. The organizers set up a conference, participants voice their opinions, notes are taken, and organizers (in some cases this includes a hired journalist and/or external readers) write the final reports. Yet, if a person's perspectives are not included in the final reporting structure and discussed in policy circles, then this calls to question the need for participants to put their names on the final reports. Stated another way, there is a prevailing "sell out" or "buy out" principle: a consensus-generating strategy inherent in these types of national conversations and meetings. In what ways do organizers foster an inclusive reporting strategy that is reflective of the array of frames and perspectives held by the participants? More broadly, people who consistently disrupt the status quo during these types of conversations may easily cease to be invited to these types of policy discussions by organizers across the country or may be sought out by particular organizations.

In a rare and highly visible example of dissent, David Ward, president of the American Council of Education and member of the nineteen-person Secretary of Education's Commission on the Future of Higher Education, was the only member of the panel who did not sign the commission's final report (U.S. Department of Education, 2006). Ward (2006) asserted that the final document established a "false sense of crisis" and blamed higher education officials and institutions for problems with multiple origins. A critical read of the commission report finds that "the text reveals an alternate agenda which is focused on maintaining and ultimately expanding hegemonic power structures and social inequalities" (Gildersleeve, Kuntz, Pasque, & Carducci, in press). Had Ward not taken the initiative to articulate his voice of dissent, this alternative perspective from within the group may have remained hidden from a wide audience. Again, power brokers matter in that they may establish processes and procedures to include alternative voices within reports, ignore such opportunities, leave dissenters to craft their own opportunities to share their perspectives, and/or alter the invitation list in the future. It is important to pay attention to and own concepts of power since spheres of influence have such a dramatic ripple effect on higher education policy and practice.

In addition to power brokers mattering, people who are not in such roles also matter. Moreover, it is important to pay attention to people without as much contextual power, since their perspectives may support the dominant perspective or offer important alternatives. The inclusion of all voices regardless of status is the best idea of democratic practice (Guarasci & Cornwell, 1997; Schoem, Hurtado, Sevig, Chesler, & Sumida, 2001). This concept of including all voices will continue to be explored further in the next two sections.

Centering Alternative and Marginalized Perspectives

Participants shared their own perspectives through intentional and public discourse. The myriad ways the participants described reality through language implicitly carries meaning (Cameron, 2001), and choice of words has tangible implications. The need to address educational inequities, particularly around issues of race, gender, and class, may be intuitive for some; however, many higher education leaders do not talk often about the topic. This study shows that when people in the field of higher education attempt to address the topic of inequality in ways that are not supported by dominant cognitive processing models, then they are often marginalized or silenced. Stated another way, ideas and concepts are being shut down based on the content (i.e., *Interconnecting and Advocacy* cognitive processing model) of the language, no matter the social identity or role of the speaker.

Tannen (1993) mentions that silencing is not necessarily connected to whether a person talks or how much a person talks. In this context, I have defined "silencing" or "to be silenced" as sharing an alternative cognitive processing model that is not picked up or incorporated into the discussion, is not incorporated into alternative models, and/or is not mentioned in the final reporting structure. In this case, it is not the lack of voices—as in silence or quiet in the room. As Marshall (1999) points out, educational policymakers utilize speech as a power tool to create power and to block, resist, or oppose particular perspectives. These conversation moves may contribute to the silencing of particular perspectives.

As emerged in this study, concepts of justice are included in all perspectives about higher education for the public good presented here. However, if the public and private goods are presented as *Interconnected* and/or the language is offered from an *Advocacy* frame, the idea is not often included in revised models for change. More specifically, *Interconnecting and Advocacy* perspectives such as Giroux and Giroux (2004), Slaughter and Rhoades (2004), and Glenn and Courtney are often blocked, redefined, and/or relegated to the margins (Also see McLaughlin & Tierney, 1993) or they need organizers who serve in important bridging positions to *give* voice to the different ideas and concepts. To give voice to someone else's idea is not necessarily a sufficient situation. To me, this is reminiscent of people stating that they "tolerate" "others" with a different perspective or identity; no one likes to be simply "tolerated" or to have someone else reiterate what they said in order for the idea to be heard.

Further, national discussions similar to the one explored here are often the genesis of educational policies. Even when a conference or discussion includes numerous people who hold alternative perspectives, these perspectives are not necessarily the frames used for crafting policy decisions. Not only do we need to include all perspectives at conferences, we need to also consider how to center these perspectives *during* discussions *and* when making decisions about access, admissions, funding, tenure, promotion, and other decisions that impact people's daily lives. Strategies that acknowledge the public and the private as interconnected, where issues of inclusion, democracy, power, access, and in/equity are directly addressed, need to be centered. It is through centering alternative perspectives as we engage in discussion and create policy change that we may interrupt the status quo. As St. John (2006) states, "it would be unjust to penalize another generation of low-income and minority students for the failures of public policy" (p. 239).

Hurtado (2007) reminds us that directly addressing inequality has been elusive in higher education. Leaders are not necessarily taught how to discuss and address these issues. Some higher education leaders who engaged in this national written and face-to-face discourse had a difficult time discussing the topic/content of educational stratification and segregation around class, race, and gender. And, as Fried (1994) points out, alternative perspectives have the strongest potential to recognize stagnant paradigms and offer new alternatives in higher education. In this sense, leaders need to adopt dialogic processes and facilitation styles that foster the expression of alternative voices and visions. Intention is not enough to make this type of concerted and equitable change in the relationships between higher education and society.

Including Voices of People from Target Identities

Colleges and universities are mired in a system and societal context that oppresses women, people of color, and people from low socioeconomic statuses and other social identities (See Adams, Bell, & Griffin, 1997; Adams, Blumenfeld, Castañeda, Hackman, Peters, & Zúñiga, 2000; Davis, 1998; hooks, 2000; hooks & West, 1991; Torres, 1998). This information, sadly, is not new. The prevailing patterns are also found during this specific conference series and supports existing literature. For example, during this conference series some men redefined what women said. The comments of some women, people of color, and graduate student were ignored. Some perspectives were directly rejected or not included in official reporting documents. Stated another way, some leaders are being relegated to the margins during the process of the conversation based, entirely or in part, on their targeted social identities (i.e., people of color, women, community partners).

To quickly reiterate, it was an academic man of color who was able to connect multiple contrasting views in a strategic manner to bridge different ideas. It was primarily people of color in marginalized positions (graduate student and community organizer) who most often acknowledged the various historical and contemporary boundaries that reproduce power relationships and that sustain cyclical oppression. In addition, it was people of color in marginalized positions (e.g., community organizer, graduate student) who initiated a change in the discussion in order to interrupt the current social reproduction of educational power in a way that echo's Bourdieu's (1986) concept of capital. There was a white male full professor who used his power to include a person of color and a woman's alternative perspectives about the ecological model (in this sense, it was not limited to people of color and women who brought out issues that did not support the dominant frames presented at the conference), and a white male organizer and white male facilitator who used their power to avoid providing alternatives to the model that was presented. Moreover, it was a woman who initially challenged the organizational model and a Latina graduate student who challenged the dominant discourse of participants. As these examples reflect, each participant, organizer, and speaker who spoke (or who chose not to speak) played an important role in shaping the discourse at this national conference series. Yet, it was the people who controlled the reporting mechanisms who had the responsibility of sorting through this discourse to determine what constituted worthiness for the final report—information sifting.

These findings reflect existing research on this topic. For example, Dijk (1993) found that certain narratives reproduced racism and social control in stories he examined about participants' neighbors. He found that these stories recreated the power relations from the narrators' position of white dominance. The findings also mirror Rex et al. (2006) who uncovered "the ways in which the dominant Eurocentric and androcentric knowledges and cultural practices delegitimize those whose race, class, ethnicity, or gender to not match the dominant discourses [in K-12 education]" (p. 17). In addition, Stanley (2006) discusses the silence of faculty of color at predominantly white colleges and universities and offers recommendations for the recruitment and retention of faculty of color in higher education.

The discounting of voices of people of color, women, and people from poor or working class families influences people's willingness and ability to continue to communicate in these types of discourse genres. On a related point, Darling and Brownlee (1984) discuss "relational acceptance" through their research on conflict management. Trust, they state, is vital.

> When the level of trust and acceptance is high, almost any effort to communicate is successful. Conversely, when the level of acceptance and trust is low, communication typically is distorted and misunderstood no matter how articulate and intelligent the parties involved in the conflict are, and responses are usually reduced to emotional and often irrational communication patterns. (pp. 249–250)

Tierney (2006b) also discusses the vital importance of trust as he considers higher education's governance and the public good through a cultural framework. He states, "trust is particularly important in organizations where risk taking needs to occur and where task requirements are not clearly delineated" (p. 193). Such is the case as leaders take risks to connect knowledge and action around critical issues toward higher education and the public good. This idea of trust is connected to both the macro- and micro-analyses presented in this study; being silenced or having ideas reframed or discounted may add to participants' low level of trust and impact participation. In addition, this reduced level of trust may be compounded by Smith's (2004) "racial battle fatigue" and Rowley's (2000) symbolic violence and social reproduction, as outlined in chapter 5. As Courtney suggests and is outlined in chapter 6, "truth telling" about historical and contemporary inequities may be one way to concretely strengthen such trust, one suggestion on *how* to address change that was overlooked by the group.

The intentional inclusion of people of color and women by the curriculum design committee was not enough to center cognitive processing models that provide an alternative to the dominant models. The same is true of the national literature on the topic, which does include some voices of people of color and women. This echoes quantitative findings from Gurin, Dey, Hurtado, and Gurin (2002) where they found that structural diversity (the number of people from diverse groups) was not enough to produce educational benefits. Instead, diverse students needed to interact with each other, not just sit in the same space and wait for important learning and democracy outcomes to occur. In a similar albeit contextually different manner, the structural diversity of the conference participants was not enough to center the perspectives of people with target identities. This is reminiscent of the ways in which desegregation of public schools has not helped educational access and equity; more than structural diversity is needed. Student diversity and inclusion are important first steps.

On a related point, people with outsider status are known to consider dominant perspectives more clearly and present innovative perspectives that have the potential to be the impetus for change (Fried, 1994). In the national policy discourse, even if higher education leaders, speakers, and authors are well-intended people (an assumption based on the stated goals of the conference series), if they continue to perpetuate dominant patterns that silence people from target identities when discussing higher education and society, important voices will be omitted from the conversation. What hope, then, do we have for making concrete national and local change around educational inequities?

Leaders need not leave it to a Latina to "turn the volume up" on her voice and take the role of the "angry woman of color," as is the case of my colleague in her research team meetings discussed in a previous chapter. The challenge for all higher education leaders is to be aware of such power dynamics, including dominant cultural values, structures, and social climates in order to create discursive spaces that center alternative perspectives and people with target identities (See Goodman, 2001). Such national discussions need to be facilitated inclusively with a clearly articulated agenda where assumptions are identified (See Brookfield, 1995).

In *Challenging Racism in Higher Education*, Chesler, Lewis, and Crowfoot (2005) identify intentional strategies for organizational change in academic settings. The authors discuss the importance of planning that tends to both incremental and radical goals for change. Further, they offer a "road to multiculturalism and justice (in primarily white colleges and universities)" (pp. 172–173) that breaks down factors for change from monocultural to transitional to multicultural organizational settings and talk about how to implement such change in the face of resistance. Only in a setting where people are cognizant of these issues and conscientious about making social change through educational equity may leaders hope to concretely change the perpetuation of the inequities in the myriad relationships between higher education and society.

Future Research

A number of studies could be pursued as an outgrowth of these particular study findings. For example, research that explores the elements of the Tricuspid Model of Interconnected Advocacy and Educational Change could uncover whether these three organizational change elements will concretely alter the relationships between higher education and society. Further, the *Public, Private, Balanced* and *Interconnected and Advocacy* frames are clearly found in the literature, but the *Advocacy* frame has not necessarily been identified and utilized as a central frame in national policy discussions. Additional research studies on the usage of this frame in the discourse, the effects of intentionally including this alternative perspective in national policy discussions and action, and the implications of this frame itself are needed. In addition, further research studies about the importance of the bridging moves that connect aspects of the various frames need to be explored.

This research study focuses on the "peak" sessions, as opposed to the majority of sessions that were "non-peak" sessions. An in-depth research study on the "non-peak" sessions is equally important in uncovering cognitive processing models toward strengthening the relationships between higher education and society. In addition, what are the various models for higher education for the public good that organizers and participants presented during the conference series? The wealth of information shared during the conference series is not collected in any one document (some of the information is collected in the final report) but may be useful for developing strategies for educational change. In addition, what are the various aspects of each of these models that do get picked up by participants and organizers?

Further, the people whom organizers include or exclude in the conferences influence the discussion. What are the ways in which people become included and excluded from the national discourse? What was an invitee's rationale for accepting or declining the invitation? What is needed is more organizational behavior research on this topic that uncovers the ways in which power brokers may use their sphere of influence to foster inclusive spaces as well as alternative strategies that interrupt the pressures of academic capitalism.

Additional research studies about the use of specific language also may be instructive. For example, a constant comparative analysis between the language used during the conferences (not limited to the findings that emerged in this research study) and the language written in the final reports may be informative. Is it representative of all participant voices? What is the comprehensive picture of what was left out? Further, the organizers' and participants' usage (and resistance to the usage) of the language of a "social movement," "covenant," and various consistently used words could be explored further. For example, the organizers define what is needed to create a social movement and then name that this conference series is a social movement.

Some participants agree that the series itself is a social movement (Courtney, Paige) and some do not (Glenn). This language of a covenant movement echoes the work of Smiley (2009) in his look at accountability and could be considered in more detail.

Added research studies on the various social identities of participants—or references to identities, including religion and age—could be quite instructive. For example, in one of the examples, Courtney questioned the group, "what sacred cows are we willing to slaughter?" This sacred cow figuratively refers to a person or thing immune to criticism or questioning. This statement alludes to the honored status of cows in Hinduism, where cows are a symbol of God's generosity to humankind. References to religiosity throughout the conference series by organizers and participants may be useful in further understanding the nuances of language in this context. In addition, exploration of social identity around chronological age is another option for further research. The curriculum design team intentionally invited people at different points in their careers. The age of a participant or organizer in relation to who is heard, or not, who speaks, or not, may also be a useful focus of research study. The concepts of chronological age and role (graduate student, postdoctoral researcher, full professor) were discussed in two member checking conversations and should be explored with greater intentionality.

The participants' narrative stories were particularly interesting. Participants' narratives about their own college experiences were more prevalent in the pilot study than in the larger research study and are explored in more detail in the former. Future research that explores the meaning making found within the personal narratives conveyed by participants throughout the conference series should be considered. Exploration of the narratives may help us to more deeply consider the myriad ways in which people make sense of their own connection in the relationships between higher education and society and provide examples of best practices across the country.

Moreover, future research studies could continue to explore the social identities of participants as connected to educational in/equity. Feminist theory was specifically utilized as a lens in chapter 6, but how do participants' and organizers' own race, gender, and class play a role in this series as considered through a critical race theory, queer, and post-Marxist perspective? How do higher education leaders interrupt current patterns of silencing people with target identities that are so pervasive in the field and in US society? What are the ways in which participants serve as allies for each other, and/or are these ways to further their own agendas? The role of allies during this conference series may help to develop the practical ways in which higher education leaders may serve as allies in future national conversations.

Hopefully, a portion of these questions will continue to be entertained by researchers and practitioners who have a desire to work toward strengthening the relationships between higher education and society in the hopes of making concrete and equitable educational change. The Tricuspid Model of Advocacy and Educational Change offered here is but one organizational strategy for higher education leaders—and leaders in various professions—as we move forward.

CHAPTER 8

Conclusion: Critical Issues and the Public Good

Judith Ramaley (2006), president of Winona State University in Minnesota and former assistant director of the Education and Human Resources Directorate at the National Science Foundation, states that the primary purposes of higher education today are

> to conduct research on important problems, ideas, and questions; to promote the application of current knowledge to societal problems; and to prepare students to address these issues through a curriculum that emphasizes scholarly work in both the liberal arts and the professions, where learning is advanced in a mode that encourages civic commitments and social responsibility. (p. 173)

Higher education needs to play an instrumental role in researching and addressing myriad issues facing the world today. Students, faculty, and staff may do this in concert *with* people in communities and organizations, where "people" is inclusive across class, nationality, race, gender, and other social identities and roles. Further, these relationships may be better served *if* we are able to communicate in ways in which we understand each other—both within groups and between groups.

This point raises questions such as the following: What happens if we truly implement organizational behavior strategies that connect knowledge and action on critical global issues as we foster the next instantiations of higher education and the public good? What happens if we actualize the Tricuspid Model for Advocacy and Educational Change in daily activities as we research and discuss social problems such as health care, global warming, poverty, educational in/equities, and the economy and—together *with* community members—advance issues toward the public good?

This book explores the perspectives of leaders as shared in the literature, speeches, and reports and behind closed doors in order to help us critically work toward inhabiting the gap between knowledge and action. More specifically, the goal of the current study was to illuminate various higher education leaders' competing frames and worldviews of higher education for the public good as found in the literature (macro-analysis) and vocalized during hours of conversation during a (twelve-day) national conference series (micro-analysis) in order to increase our understanding of the perspectives and knowledge and to translate this knowledge into action.

This particular conference series is a vent that enables us to consider the entire field of higher education, both within and outside the brick and mortar of the college campus. In fact, the conference series explored in this study and organizations that co-sponsored it are examples of people working within the field of higher education attempting to make needed change in the national dialogue. For example, the conference series curriculum committee tried intentionally to include people not traditionally invited to this type of an event and to invite a diverse (across race, gender, and position) group of participants. The importance of this study focuses on how—even with the intention of and action toward making change in the field as is evidenced by (1) exploring this topic with higher education leaders through conferences, (2) providing funding for participation across the invitees, and (3) inviting people to discuss the problem—issues of power, educational access and in/equity, and marginalization of ideas and people remain and need to be addressed by leaders.

As noted in the acknowledgments, I am extremely thankful that the people within this organization and cosponsoring organizations took the initiative to gather higher education leaders who, in discussing this topic, were willing to open themselves up to this type of unintended scrutiny in order to continue to learn from one another. To me, this effort is a solid step toward "inhabiting the gap" between knowledge and action in the hopes of making change.

What emerged from the discourse is a multilevel problem that is consistent throughout the literature review, pilot study, and current conference series with dominant cognitive processing models. This multilevel problem manifests when people utilize the same words to mean different things (i.e., social capital has various meanings as described in the literature review analysis), alternative perspectives are silenced (i.e., in the final reporting structures of a national convening), and people of color and women are relegated to the margins of discussions (i.e., ignoring and redefining a request for information).

Moreover, the inherent power of organizers, funders, and facilitators cannot be overstated. If such a model for process as the Tricuspid Model for Advocacy and Change was employed, alternative action strategies in terms of *how* to make change in higher education may not have been overlooked in the discourse, such as Judith's idiosyncratic method for educational change, Stephanie's grassroots models, and Courtney's truth-telling examples to develop trust between communities (see chapter 6). These strategies would have been centered and provide alternative options, in addition to existing options, to connect knowledge and action as leaders address critical issues for higher education and the public good.

Each leader's perspective has a different set of ideas, assumptions, and implications for the continuation or interruption of current paradigms in research and policy about higher education for the public good (see Bolman & Deal, 2008). By understanding more about various cognitive processing models—or frames—of higher education's relationship with society, and the tensions created between these frames, we are able to see more of the perspectives and make more informed choices about how to work toward systemic and equitable change.

As I have argued, higher education leaders cannot afford to be complacent in this climate of educational inequity and let dominant arguments limited to the private/economic good of higher education prevail. Uncovering various visions of higher education's relationships to society is paramount during the current period of dramatic change in higher education. If a more thorough understanding of myriad public good perspectives is not offered, then dominant communicative models shared in academic discourse genres may continue to perpetuate the current ideas of higher education's relationship with society—from an economic rationalization perspective—without consideration of alternative perspectives.

As Giroux and Giroux (2004) argue, to silo public and private good arguments perpetuates systemic oppression and stratification. More specifically, a private good argument addresses issues of economic injustice but does not connect this injustice and action with historic and contemporary cyclical oppression as identified by Courtney in the face-to-face dialogue. A public good argument void of the economic discussion ignores the inherent power of the market. The *Interconnected and Advocacy* perspective proposed by leaders (and also marginalized in the discourse)—where the public and private good are understood as interconnected—is one that may address cyclical oppression around race, class, nationality, and gender and work toward interrupting the current paradigms as it not only pays attention to both the public and the private but also discusses the ways in which they influence each other to perpetuate or change oppression.

Further, the interconnections between the public and private goods from a critical inquiry perspective focus on an emancipatory agenda for higher education and society. From this perspective it is particularly important for leaders to initiate change in order to address educational inequities across race, gender, and class within and outside of the system of higher education as the contemporary public agenda discourse actively works to perpetuate dominant paradigms of stratification (Gildersleeve, Kuntz, Pasque, & Carducci, in press). Change to actualize a true and inclusive democracy—inclusive of both dominant and marginalized voices—is central to the future of higher education.

This begs the question, what is the role of higher education leaders—inside and outside of exclusive environments—as we reconceptualize critical issues for higher education and the public good with regard to the public and private benefits of college? The lines between these historically separated entities are blurred and leaders must recognize that the false dichotomy of public and private has historically perpetuated racism, sexism, and classism. The current, significant concentration on market values by higher education leaders stratifies the "haves" and the "have-nots" thereby decreasing the value of higher education as a public good. It is up to higher education leaders to make this shift—from a focus on the private benefits of higher education to a complex understanding of the interconnection between public and private benefits. This argument is echoed by a number of other scholars (Bowen & Bok, 1998; Brint & Karabel, 1989; Green & Trent, 2005; Hagedorn & Tierney, 2002; Labaree, 1997; 2007) and becomes even more important during these critical political, cultural, and economic times. A university might "open up a space for more than just a democracy" (Giroux & Giroux, 2004, p. 213) including the potential to address large societal and global problems and create spaces for dialogue and debate.

It must also be noted that I do not assume that all leaders are conscious of the ways in which they/we perpetuate the status quo as the field of higher education often does not focus on educating ourselves and students about these issues. Leaders' dismissal or exclusion of marginalized voices, or inclusion of change strategies that are equitable for some but not all people in our communities, may not be intentional. As Hurtado (2007) reminds us, "addressing inequality in American society, however, has been elusive in higher education and absent from the nation's agenda" (p. 186). This is where I focus my approach—higher education leaders need to pay more attention to the ways in which we are perpetuating or interrupting the status quo that is not functioning effectively. Intention is not enough; higher education leaders need to speak dangerous truths as these discourses have the potential to impact policy, procedures, and—in turn—people's daily lives.

PARRHESIASTES OF THE *AGORA*: SPEAKERS OF DANGEROUS TRUTHS

As described in chapter 2 of this book, Aristotle was one of the few classical thinkers about democracy to explicitly talk about three spheres of human activities, the *oikos* (private), the *ekklēsia* (public), and the *agora* (the overlapping of the *oikos* and *ekklēsia*) (as cited in Castoriadis, 1997). This important overlap or interconnection of both private and public spheres *simultaneously* pulls forward the strong points of various ideas and strengthens arguments for effective policy and action. Taken together, these perspectives acknowledge the importance of the market *and* address inequities around race, class, and gender and other critical social issues in the global society. As outlined throughout this text, there are a number of scholars who speak to this intersection; however, the content of what these scholars say and the process of how they say it continue to be marginalized when it comes to national policy circles. Such scholars in ancient Greece were called *parrhesiastes*, the speakers of dangerous truths (Foucault w/ Pearson, 2001). See figure 8.1.

Bourdieu provides language useful for modern day *parrhesiastic* leaders to utilize as we work to connect public benefits together with an economic benefits approach. Specifically, Bourdieu's (1986) definition of social capital connects three sources of capital (economic, cultural, and social) in order to create an aggregate of resources linked to a network of relationships. He defines social capital as grounded in theories of symbolic power and social reproduction where social capital is a tool of reproduction for the privileged. In this sense, social, cultural, and economic capital continue to be reproduced in a way that supports the privileged and marginalizes the oppressed. From this perspective, all things are political, including the exclusion/inclusion of the public benefits of higher education in higher education's public agenda. The recognition of various sources of capital becomes instrumental as leaders choose what to reproduce and how to effect change.

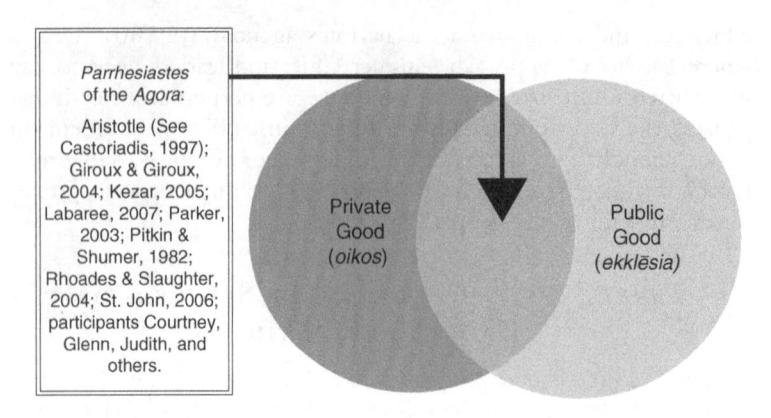

Figure 8.1 *Parrhesiastes* of the *Agora*: Speakers of Dangerous Truths of the Interconnections of Public and Private

Also instructive in these efforts that utilize the Tricuspid Model for organizational and societal change (presented in the previous chapter) are Harro's Cycle of Socialization (2000a) and Cycle of Liberation (2000b). Harro (2000a) describes the Cycle of Socialization process as "*pervasive* (coming from all sides and sources), *consistent* (patterned and predictable), *circular* (self-supporting), *self-perpetuating* (intra-dependent) and often *invisible* (unconscious and unmanned)" (p. 15). Specifically, we are born into a preexisting world and socialized with rules, norms, and expectations. These perspectives are reinforced through institutional and cultural messages that result in dissonance, silence, collusion, ignorance, violence, and internalized patterns of power. The Cycle of Socialization offers two options: (1) do nothing and perpetuate the status quo, or (2) make change by raising consciousness, interrupting patterns, educating one's self and others, and taking action.

Harro (2000b) also offers a Cycle of Liberation regarding individual, collaborative, community, and culture change. Change, in this instance, requires a "struggle against discrimination based on race, class, gender, sexual identity, ableism and age—those barriers that keep large portions of the population from having access to economic and social justice, from being able to participate fully in the decisions affecting our lives, from having a full share of both the rights and responsibilities of living in a free society" (p. 450). Such change interrupts dominant paradigms of oppression through the use of transformational and collaborative relationships. This cycle begins on the intrapersonal level, where social inequities create cognitive dissonance when people cannot rectify in their own minds what they see happening versus what they think should be happening to address critical social issues. From here, a person gets "ready" by empowering him/herself to dismantle collusion, privilege, and internalized oppression. At this point, the person reaches out toward others, seeks experiences, names injustices, and uses tools for change. Next is the interpersonal phase in which people build community. In this work with others, we seek people "like us" for support and people "different from us" for building coalitions and questioning assumptions in structures and systems (p. 464). Coalescing is an important stage where people engage in organizing, action planning, fundraising, educating, renaming, learning about being an ally, and moving into action. Creating such systemic change critically transforms institutions and creates a new culture through influencing policy, structures, leadership, and a shared sense of power. This work is maintained as additional systemic changes are initiated.

A third instructive framework offered is from Owens (2009) who discusses dimensions of the relevance of privilege for higher education leaders interested in diversity leadership at predominantly white institutions. The dimensions of power, perception/symbolism, consistency with diversity objectives, and effectiveness are useful for leaders to consider in various higher education contexts. Further, Cullen (2008) analyzes the "35 dumb things well-intended people say" and offers leaders core concepts and steps for lessening the diversity gap.

In this sense, Aristotle, Harro, Owens, and Cullen (and many others) provide instructive frameworks for modern-day *parrhesiastes* to utilize when addressing educational inequities in higher education while connecting notions of private and public goods. It is up to leaders to seek and implement various strategies for change on a continual basis.

In sum, higher education leaders have been called upon to change the current trajectory that perpetuates educational stratification and academic capitalism (Gildersleeve, Kuntz, Pasque, & Carducci, in press; Giroux & Giroux, 2004; Rhoades & Slaughter, 2004). This change can be enhanced through strategic relationships between community-university partners as we work toward social change with deep and/or broad impact. In addition, the *Interconnected /Advocacy* perspective is imperative as a foundational element in higher education's public agenda as it both blends and connects multiple perspectives. This complex perspective is inclusive of the familiar economic/ private good perspective as it considers the private good's interconnections with the public goods perspective.

Further, this perspective *specifically names* justice and equity as crucial to strengthening the relationships between higher education and society. Notably, "Equity" is intentionally used instead of "equality" as equality may not be attainable until equity is actualized. Justice and equity across race, gender, and class are considered simultaneously through the interconnections of public and private goods of education. It is the specificity of naming social justice and equity "for whom" (inclusive of all people) and "in what ways may this be possible" (for equity across race, class, nationality, gender, etc.), while addressing historical and contemporary oppression, that distinguishes this frame's concepts about social justice and advocacy.

Not only does this frame specifically name justice and equity as important, it also urges higher education leaders to reject the limitations of a singular approach and encourages an interconnected approach in order to break the cycle of *cumulative oppression* (Pasque & Rex, 2010) across generations. Inclusive organizational behavior practices by leaders will help center important perspectives that may have the potential to affect critical change.

Higher education is in need of pervasive leadership by *parrhesiastes* of the *agora* to speak dangerous truths in order to initiate educational change. Such pervasive leadership has been defined as

> individually generated relationships and actions among members throughout an organization focused on struggling together to influence and promote organizational learning and accomplish positive changes to benefit the common good. (Love & Estanek, 2004, p. 38)

Pervasive leadership that acknowledges power and creates strategies for equitable change in order to address complex problems is vital to social transformation. It not only recognizes the importance of power brokers but also centers voices of all people and includes marginalized perspectives. In addition, it centers the content of alternative paradigms in the hopes of transforming gridlock into action as we craft the new future of higher education and the public good.

Notes

1 Introduction to the Contemporary Context

1. These statistics do not include information about Native American or Asian American students. In addition, they do not break down the statistics within racial and ethnic groups that uncover further disparity within and between student populations.

2 Higher Education for the Public Good: A Typology

1. Joe Luft, and Harry Ingham (Luft, 1970) created the Johary Window to describe self-disclosure. The four quadrants, or sections of a window separated by a movable pane, are open, hidden, blind, and unknown. The open quadrant is information that a person knows about him/herself and other people know about that person. The hidden quadrant is what a person knows about him/herself and yet does not reveal to others. The blind quadrant is information known to others and not to him/herself. The unknown quadrant is information not known to the person him/herself or to others.
2. These adapted figures and a version of this chapter have been shared and discussed with Jamie Merisotis, president of the Institute for Higher Education Policy at the time the Array of Benefits was published.
3. "His/her" is currently being referred to as "hir" in the field of student affairs and beyond. I use "hir" in this volume as it is inclusive of males, females, and transgendered persons.
4. Tradeoffs are one of the foundational principles of economic theory (Buchanan & Tullock, 1962).

3 Research Design

1. This is distinguishable from social identity development theories that often define identity as a process where a person moves from one stage to another (See Evans, Forney, & Guido-DiBrito, 1998).
2. The final reports are treated as data for this study instead of an article for a literature review. As such, and for the purposes of confidentiality, these reports are not included in the reference list.

3. The small groups did not report out in this large conference of over 200 people. Instead, ideas were compiled, written down, and passed out to the large group. Then, one delegate from each small group facilitated a conversation at tables of approximately 10 people in order to discuss the compilation document and ask questions. At this point, there was a large group facilitated discussion that was mainly a series of questions posed by participants and answers given by the main speaker, but it did not yield a discussion between participants.

4 BEHIND CLOSED DOORS: COMMUNICATION THEORIES AND CONVERSATION ANALYSIS

1. I have added gender expression to the Hardiman and Jackson (1997) definition of social identity to communicate the explicit differences between gender, sexual orientation, and gender expression.
2. It is important to note that not all code-switching was verbalized by African American participants during this conference series. For example, in the first conference a white woman participant switched from a conversational, discussion style of speech to a more of a lecture style when sharing the history of higher education as related to the discussion.

5 BEHIND CLOSED DOORS: A CRITICAL ANALYSIS OF THE DISCOURSE

1. Trenholm and Jensen (1992) title this the "relational level" of the conversation and their definition includes both verbal and nonverbal communication.
2. The pilot study explored a different conference in the same series.
3. Each participant in this study was provided with a pseudonym for the purposes of analysis. Where deemed appropriate, an attempt was made to provide consistency between participants' names and pseudonyms in relation to social identity as defined by Adams (1997). Specifically, gender, race, ethnic, religious, and historical contexts were considered. The website Behindthename.com (2006), which provides the origin and definition of many names, was consulted during this process.
4. For an analysis of when the model was questioned during other conferences in this series, see the next chapter.
5. This is not to negate the important information shared in these reports that emerged from the "non-peak" conversations.
6. These statistics do not address the disparities between all groups of people of color or give a breakdown of the data for each racial and ethnic group that can uncover further disparity between student populations.

7. The woman who shared her experiences with me on a number of occasions was shown this section of the analysis. I have her permission to use this story.
8. Smith does footnote that he is not equating the daily experiences of people in the military with those of faculty of color but suggests this as a useful metaphor.

References

Abes, E. A. (2009). Theoretical borderlands: Using multiple theoretical perspectives to challenge inequitable power structures in student development theory. *Journal of College Student Development. 50*(2). 141–156.

Abes, E. A. & Kasch, D. (2007). Using queer theory to explore lesbian college students' multiple dimensions of identity. *Journal of College Student Development. 48*(6). 619–636.

Adams, M., Bell, L. A., & Griffin P. (1997). *Teaching for diversity and social justice: A sourcebook.* New York: Routledge.

Adams, M., Blumenfeld, W. J., Castañeda, R., Hackman, H. W., Peters, M. L., & Zúñiga, Z. (2000). *Readings for diversity and social justice: An anthology on racism, antisemitism, sexism, heterosexism, ableism, and classism.* New York: Routledge.

Allen, E. J., Iverson, S. Ropers-Huilman, R. (Eds.). (2009). *Reconstructing policy in higher education: Feminist poststructural perspectives.* Clifton: Routledge.

American Association of Colleges and Universities. (2002). *Greater expectations: A new vision for learning as a nation goes to college.* Washington, DC: Author.

American Association of Colleges and Universities. (2006). *Liberal education and America's promise.* Retrieved on March 8, 2006 from http://www.aacu.org/advocacy.

American Council on Education. (2002). *Minorities in higher education, 2001–2002. Nineteenth annual status report.* Washington, DC: Author.

American Council on Education. (2006). *Solutions for our future.* Retrieved March 8, 2006, from http://www.solutionsforourfuture.org/site/PageServer.

American Federation of Teachers, Higher Education. (March 2009). Generating power: Mobilizing the union to revitalize higher education. Miami Beach, FL.

Apple, M. (2006). *Educating the "right" way: Markets, standards, God, and inequality* (2nd ed.). New York: Routledge.

Arminio, J. L., & Hultgren, F. H. (2002). Breaking out from the shadow: The question of criteria in qualitative research. *Journal of College Student Development. 43*(4), 446–460.

Association for the Study of Higher Education. (2006). *The social responsibility of higher education.* Philadelphia: Author.

Astin, H. S. & Leland, C. (1991). *Women of influence, women of vision: A cross-generational study of leaders and social change.* San Francisco: Jossey-Bass.

Astin, A. W., Vogelgesang, L. J., Ikeda, E. K., & Yee, J. A. (2000). *How service learning affects students.* Los Angeles: Higher Education Research Institute.

Bakhtin, M. M. (1981). *The dialogic imagination.* Austin: University of Texas Press.

Ball, S. (1990). *Politics and policy making in education: Explorations in policy sociology.* London: Routledge.

Barash, D. P. (2004, April 16). Caught between choices: Personal gain vs. public good. *The Chronicle of Higher Education,* pp. B12–14.

Barber, B. (1998). *A passion for democracy.* Princeton: Princeton University Press.

Bartik, T. J. (2004). *Increasing the economic development benefits of higher education in Michigan* (No. 04–106). Kalamazoo: Upjohn Institute.

Battiste, M. (2006, May). The global challenge: Research ethics for protecting Indigenous knowledge and heritage. Keynote address at the Congress of Qualitative Inquiry. Urbana-Champagne, IL, USA.

Baum, S., & Payea, K. (2004). *Education pays 2004: The benefits of education for individuals and society.* New York: College Board.

Bauman, Z. (1999). *In search of politics.* Cambridge: Polity Press.

Bautista, M. N. M. (2006). *Appearance, subculture, and narrative identity in punks, darks, and rockabillies in Mexico City.* Paper presented at the Congress of Qualitative Inquiry, Urbana-Champagne, IL.

Becker, G. S. (1964/1993). *Human capital: A theoretical and empirical analysis with special reference to education* (3rd ed.). Chicago: University of Chicago Press.

Becker, W., & Lewis, D. (1993). *Higher education and economic growth.* Norwell: Kluwer Academic Publishers.

Behind the Name. (2006). Retrieved on June 8, 2006 from http://www.behindthename.com/.

Behling, C. (2007). *Facilitator training for intergroup dialogues on race, gender, class, and religion.* Paper presented at the Intergroup relations retreat for practicum facilitators, Ann Arbor, MI.

Bell, L. A. (1997). Theoretical foundations for social justice education. In B. Adams, L. A. Bell & P. Griffin (Eds.). *Teaching for diversity and social justice.* New York: Routledge.

Bem, S. L. (1993). *The lenses of gender: Transforming the debate on sexual inequality.* New Haven: Yale University Press.

Bensimon, E. M. & Marshall, C. (2003). Like it or not: Feminist critical policy matters. *Journal of Higher Education.* 74(3). 337–349.

Berzon, B. (1996). *Setting them straight: You can do something about bigotry and homophobia in your life.* New York: Penguin.

Blinder, A. S., & Weiss, Y. (1976). Human capital and labor supply: A synthesis. *The Journal of Political Economy, 84*(3), 449–472.

Bloom, A. (1987). *The closing of the American mind: How higher education has failed democracy and impoverished the souls of today's students*. New York: Simon & Schuster.

Bloom, D., Hartley, M., & Rosovsky, H. (2006). Beyond private gain: The public benefits of higher education. In P. G. Altbach & J. Forrest (Eds.). *International Handbook of Higher Education*. Netherlands: Springer.

Blumenfeld, W. J., & Raymond, D. (2000). Prejudice and discrimination. In M. Adams, W. J. Blumenfeld, R. Castañeda, H. W. Hackman, M. L. Peters, & X. Zúñiga (Eds.). *Readings for diversity and social justice*. New York: Routledge.

Bok, D. (1982). *Beyond the ivory tower: Social responsibilities of the modern university*. Cambridge: Harvard University Press.

———. (2003). *Universities in the marketplace: The commercialization of higher education*. Princeton: Princeton University Press.

Bolman, T. E. & Deal, L. G. (2008). *Reframing organizations: Artistry, choice and leadership* (4th ed.). San Francisco: Jossey-Bass.

Boulus, M. A. (2003). *Challenges and implications: Declining state support of Michigan public higher education*. Lansing: Presidents Council State Universities of Michigan.

Bourdieu, P. (1986). The forms of capital. In J. Richardson (Ed.). *Handbook of theory and research for the sociology of education*, pp. 241–258. Westport: Greenwood Press.

Bowen, H. R. (1977). *Investment in learning: The individual and social value of American higher education*. San Francisco: Jossey-Bass.

Bowen, W. G., & Bok, D. (1998). *The shape of the river: Long-term consequences of considering race in college and university admissions*. Princeton: Princeton University Press.

Brandl, J., & Holdsworth, J. M. (2003). On measuring what universities do: A reprise. In D. R. Lewis & J. Hearn (Eds.). *The public research university: Serving the public good in new times*. New York: University Press of America.

Brandl, J., & Weber, V. (1995). *An agenda for reform: Competition, community, concentration*. Minneapolis: Office of the Governor.

Bringle, R. G. & Hatcher, J. A. (2002). Campus-community partnerships: The terms of engagement. *Journal of Social Issues, 58*, 503–516.

Brint, S. & Karabel, J. (1989). *The diverted dream: Community colleges and the promise of educational opportunity in America, 1900–1985*. New York: Oxford University Press.

Brookfield, S. D. (1995). *Becoming a critically reflective teacher*. San Francisco: Jossey-Bass.

Brothen, T., & Wambach, C. (2004). A historical note on retention: The founding of general college. In I. M. Duranczyk, J. L. Higbee, & D. B. Lundell (Eds.). *Best practices for access and retention in higher education*. Minneapolis: Center for Research on Developmental Education and Urban Literacy.

Brown, L., & Strega, S. (2005). Transgressive possibilities. In L. Brown & S. Strega (Eds.). *Research as resistance: Critical, indigenous, and anti-oppressive approaches,* pp. 1–17. Toronto: Canadian Scholars' Press.

Brown, P., & Levinson, S. C. (1987). *Politeness: Some universals in language usage.* Cambridge: Cambridge University Press.

Brown, R., & Schubert, J. D. (2000). *Knowledge and power in higher education: A reader.* New York: Teachers College Press.

Bruner, J. S. (1990). *Acts of meaning.* Cambridge: Harvard University Press.

Buchanan, J. M., & Tullock, G. (1962). *The calculus of consent: Logical foundations of constitutional democracy.* Ann Arbor: University of Michigan Press.

Burke, J. C. (2005). Preface. In J. C. Burke (Ed.). *Achieving accountability in higher education: Balancing public, academic, and market demands.* ix–xix. San Francisco: Jossey-Bass.

Burke Johnson, R. (1997). Examining the validity structure of qualitative research. *Education, 118*(2), 282–292.

Butler, J. (1990/1999). *Gender trouble: Feminism and the subversion of identity.* New York: Routledge.

Cage, M. C. (1991, June 26). Thirty states cut higher-education budgets by an average of 3.9% in fiscal 90–91. *The Chronicle of Higher Education,* pp. A1–A2.

Callan, P. M., & Finney, J. E. (2002, July/August). Assessing educational capital: An imperative for policy. *Change, 34*(4), 25–31.

Cameron, D. (2001). *Working with spoken discourse.* Thousand Oaks: Sage.

Campus Compact. (2004). *National coalition.* Retrieved November 12, 2004 from www.campuscompact.org

Campus Compact & American Association of Colleges and Universities. (2006). *Center for liberal education and civic engagement initiatives.* Retrieved March 8, 2006 from http://www.compact.org/clece/detail.php?id=4

Cantor, N. (2003). *Celebration of diversity: A call for action keynote address.* Retrieved December 2, 2003, from http://www.oc.uiuc.edu/chancellor/cantor11_19.htm

———. (2006). *About Chancellor Cantor.* Retrieved on August 5, 2009 from http://www.syr.edu/chancellor/about/index.html.

———. (2007). *The emerging challenges to higher education.* Remarks presented at the Center for the Study of Higher and Postsecondary Education 50th Anniversary Inaugural Event, Ann Arbor, MI.

Carnevale, A. P. & Fry, R. A. (2001). *Economics, demography, and the future of higher education policy.* New York: National Governors' Association.

Castoriadis, C. (1997). Democracy as procedure and democracy as regime. *Constellations, 4*(1), 1–18.

Charmaz, K. (2005). Grounded theory in the 21st century: Applications for advancing social justice theory. *The Sage Handbook of Qualitative Research* (3rd ed.), pp. 507–535. Thousand Oaks: Sage.

Chase, S. E. (2005). Narrative inquiry: Multiple lenses, approaches, voices. In N. Denzin & Y. Lincoln (Eds.). *The sage handbook of qualitative research* (3rd ed.), pp. 651–679. Thousand Oaks: Sage.

Chesler, M., Lewis, A., & Crowfoot, J. (2005). *Challenging racism in higher education.* New York: Rowman & Littlefield.

Chesler, M., Peet, M., & Sevig, T. (2003). Blinded by whiteness: The development of white students' racial awareness. In E. Bonilla-Silva & E. Doane (Eds.). *White out: The continuing significance of racism,* pp. 215–230. New York: Routledge.

Coleman, J. (1988). Social capital in the creation of human capital. *American Journal of Sociology, 94*(Issue Supplement), S95–S120.

———. (1992). Some points on choice in education. *Sociology of Education, 65*(4), 260–262.

Coulter, C. A., & Smith, M. L. (2009). The construction zone: Literary elements in narrative research. *Educational Researcher, 38*(8), 577–590.

Council of Graduate Schools. (2008). *Making a difference: A selection of graduate degree holders and their contributions to the public good.* Publication presented at the CGS Legislative Forum, April 24, 2008 in Washington, DC.

Cremin, L. A. (1988). *American education: The metropolitan experience, 1876–1880.* New York: Harper & Row.

Crensaw, K. W. (2002). The first decade: Critical reflections, or "a foot in the closing door". In F. Valdes, Culp, J. M., & A. P. Harris, (Eds.). *Crossroads, directions and a new critical race theory,* pp. 9–31. Philadelphia: Temple University Press.

Creswell, J. W. (2003). *Research design: Qualitative, quantitative, and mixed methods approaches* (2nd ed.). Thousand Oaks: Sage.

Cullen, M. (2008). *35 dumb things well-intended people say: Surprising things we say that widen the diversity gap.* Garden City: Morgan-James.

Currie, J., & Newson, J. (1998). *Universities and globalization: Critical perspectives.* Thousand Oaks: Sage.

Daiute, C. (2004). Creative uses of cultural genres. In C. Daiute & C. Lightfoot (Eds.). *Narrative analysis: Studying the development of individuals in society,* pp. 111–134. Thousand Oaks: Sage.

Daiute, C. & Lightfoot, C. (2004). Theory and craft in narrative inquiry. In C. Daiute & C. Lightfoot (Eds.). *Narrative analysis: Studying the development of individuals in society,* pp. viii–xviii. Thousand Oaks: Sage.

David, L., Bender, L., & Burns, S. (Producers) & Guggenheim, D. (Director). (2006). *An inconvenient truth* [Motion picture]. United States: Paramount Classics.

Davis, A. (1998). *The Angela Y. Davis reader.* Malden, MA: Blackwell.

Day, J. C., & Newburger, E. C. (2002). *The big payoff: Educational attainment and synthetic estimates of work-life.* Washington, DC: US Department of Commerce, Economics, and Statistics Administration, US Census Bureau.

Decker, P. T. (1997). *Findings from Education and Economy: An Indicators Report*. Washington, DC: National Center for Education Statistics.

Delgado, R. & Stefancic, J. (Eds.). (2002). *Critical race theory: The cutting edge*. Philadelphia: Temple University Press.

Denzin, N. K. (1978). *Sociological methods: A sourcebook*. New York: McGraw-Hill.

Denzin, N. K. & Giardina, M. D. (2008). *Qualitative inquiry and the politics of evidence*. Walnut Creek, CA: Left Coast Press.

————. (2009). Qualitative inquiry and social justice: Toward a politics of hope. In N. K. Denzin & M. D. Giardina (Eds.). *Qualitative inquiry and social justice*. pp. 11–52. Walnut Creek, CA: Left Coast Press.

DesJardins, S. L. (2003). The monetary returns to instruction. In D. R. Lewis & J. Hearn (Eds.). *The public research university: Serving the public good in new times*. New York: University Press of America.

Deveare Smith, A. (1994). *Twilight—Los Angeles, 1992 on the road: A search for American character*. New York: Anchor Books.

DeVito, J. A. (1992). *The interpersonal communication book* (6th ed.). New York: Harper Collins.

Dewey, J. (1916). *Democracy and education*. New York: Macmillan.

Dews C. L. & Law, C. L. (1995). *This fine place so far from home: Voices of academics from the working class*. Philadelphia: Temple University Press.

Dijk, T. A. V. (1993). Stories and racism. In D. K. Mumby (Ed.). *Narrative and social control: Critical perspectives*, pp. 121–142. Newbury Park: Sage.

Dika, S. L. & Singh, K. (2002). Applications of social capital in educational literature: A critical synthesis. *Review of Educational Research, 72*(1), 31–60.

D'Souza, D. (1991). *Illiberal Education*. New York: Free Press.

Edwards, D. & Potter, J. (1992). *Discursive psychology*. London: Sage.

Eisenhart, M. & Howe, K. (1992). Validity in educational research. In M. LeCompte, W. Millroy, & J. Preissle (Eds.). *The handbook of qualitative research in education*, pp. 643–680. New York: Academic Press.

Erb, R. (June 6, 2009). Why don't colleges cut costs, tuition? *Detroit Free Press*. (1A, 4A).

Erickson, F. (2004). *Talk and social theory: Ecologies of speaking and listening in everyday life*. Cambridge: Polity Press.

Evans, N. J., Forney, D. S., Guido-DiBrito, F. (1998). *Student development in college: Theory, Research and Practice*. San Francisco: Jossey-Bass.

Fairclough, N. (2001). The discourse of new labour: Critical discourse analysis. In M. Wetherell, S. Taylor, & S. J. Yates (Eds.). *Discourse as data: A guide for analysis*. Thousand Oaks: Sage.

Fine, M. (2000). Dis-stance and other stances: Negotiations of power inside feminist research. In J. Glazer-Raymo, B. K. Townsend, & B. Ropers-Huilman (Eds.). *Women in higher education: A feminist perspective* (pp. 117–132). Boston: Pearson.

Fiorito, R., & Kollintzas, T. (2004). Public goods, merit goods, and the relation between private and government consumption. *European Economic Review, 48*(6), 1367.

Fitzgerald, H. E., Zimmerman, D. L., Burack, C. & Siefer, S. (Eds.). (2010). *Handbook of engaged scholarship: Contemporary landscapes, future directions: Volume II: Community-campus partnerships.* Lansing: Michigan State University Press.

Foley, D. (1990). *Learning capitalist culture: Deep in the heart of Tejas.* Philadelphia: University of Pennsylvania Press.

Foucault, M. (1976). *The archaeology of knowledge.* New York: Harper & Row.

Foucault, M. (author) & Pearson, J. (Ed.). (2001). *Michel Foucault: Fearless speech.* Los Angeles: Semiotext(e).

Frank, R. H. (2005, February 17). If firmly believed, the theory that self-interest is the sole motivator appears to be self-fulfilling. *The New York Times,* p. C2.

Freire, P. (1970/2002). *Pedagogy of the oppressed* (30th anniversary ed.). New York: Continuum.

———. (1973). *Education for critical consciousness.* New York: Seabury Press.

Fried, J. (1994). In groups, out groups, paradigms, and perceptions. In J. Fried *Different Voices: Gender and Perspective in Student Affairs Administration,* pp. 30–45. Washington, DC: National Association of Student Personnel Administrators.

Friedman, M. & Friedman, R. (1980). *Free to choose: A personal statement.* New York: Harcourt Brace Jovanovich.

Galston, W. A. (2001). Political knowledge, political engagement, and civic education. *Annual Review of Political Science, 4,* 217–234.

Galura, J. A., Pasque, P. A., Schoem, D. & Howard, J. (Eds.). (2004). *Engaging the whole of service-learning, diversity, and learning communities.* Ann Arbor: OCSL Press.

Gándara, P. (2002). Meeting common goals: Linking K-12 and college interventions. In W. G. Tierney & L. S. Hagedorn (Eds.). *Increasing access to college: Extending possibilities for all students.* Albany: State University of New York Press.

Gee, J. P. (2005). *An introduction to discourse analysis: Theory and method* (2nd ed.). New York: Routledge.

Geiger, R. (1999). The ten generations of American higher education. In P. G. Altbach, R. O. Berdahl, & P. J. Gumport (Eds.). *American higher education in the twenty-first century.* Baltimore: Johns Hopkins University Press.

Gildersleeve, R. E., Kuntz, A., Pasque, P. A., & Carducci, R. (in press). The role of critical inquiry in (re)constructing the public agenda for higher education: Confronting the conservative modernization of the academy. *The Review of Higher Education.*

Gilligan, C. (1982). *In a different voice: Psychological theory and women's development.* Cambridge: Harvard University Press.

———. (1987). Moral orientation and moral development. In E. F. Kittay & D. T. Meyers (Eds.). *Woman and moral theory,* pp. 19–33. Totowa: Rowman & Littlefield.

———. (1988). Two moral orientations: Gender differences and similarities. *Merrill-Palmber Quarterly, 34*(3), 223–237.

Gilligan, C., Rogers, A. G., & Tolman, D. L. (1991). *Women, girls and psychotherapy: Framing resistance.* New York: Haworth.

Giroux, H. A. (2001). *Theory and resistance in education: Towards a pedagogy for the opposition.* Westport: Bergin & Garvey.

Giroux, H. A. & Giroux, S. S. (2004). *Take back higher education: Race, youth, and the crisis of democracy in the post-civil rights era.* New York: Palgrave Macmillan.

Goffman, E. (1981). *Forms of talk.* Philadelphia: University of Pennsylvania Press.

González, K. P. & Padilla, R. V. (2008). Latina/o faculty perspectives on higher education for the public good: An intergenerational approach. In K. P. González & R. V. Padilla (Eds.). *Doing the public good: Latina/o scholars engage in civic participation,* pp. 1–12. Sterling, VA: Stylus.

Goodman, D. J. (2001). *Promoting diversity and social justice: Educating people from privilege groups.* Thousand Oaks: Sage.

Gottlieb, P. D., & Fogarty, M. (2003). Educational attainment and metropolitan growth. *Economic Development Quarterly, 17*(4), 325–336.

Gratz v. Bollinger, 123 S.Ct. 2411 (2003).

Green, D. O., & Trent, W. (2005). The public good and a racially diverse democracy. In A. J. Kezar, T. C. Chambers, & J. Burkhardt (Eds.). *Higher education for the public good: Emerging voices from a national movement.* San Francisco: Jossey-Bass.

Gronbeck, B. E., McKerrow, R. E., Ehninger, D., & Monroe, A. H. (1990). *Principles and types of speech communication* (11th ed.). Glenview, IL: Scott, Foresman and Company.

Grutter v. Bollinger, 123 S.Ct. 2325 (2003).

Guarasci, R. (2009). *About Wagner; Presidents biography.* Retrieved on August 5, 2009 from http://www.wagner.edu/about_wagner/presidents_bio

Guarasci, R. & Cornwell, G. H. (1997). *Democratic education in an age of difference: Redefining citizenship in higher education.* San Francisco: Jossey-Bass.

Gumperz, J. (1982). *Language and social identity.* Cambridge: Cambridge University Press.

Gurin, P., Dey, E. L., Hurtado, S., & Gurin, G. (2002). Diversity and higher education: Theory and impact on educational outcomes. *Harvard Educational Review, 72*(3), 330–366.

Gutmann, A. (1999). *Democratic education.* Princeton: Princeton University Press.

Gutmann, A. (2008). *Office of the president: Biography.* Retrieved on August 5, 2009 from http://www.upenn.edu/president/gutmann/biography.html

Habermas, J. (1971). *Knowledge and human interests.* Boston: Beacon Press.

Hagedorn, L. S., & Tierney, W. G. (2002). Cultural capital and the struggle for educational equity. In W. G. Tierney & L. S. Hagedorn (Eds.). *Increasing access to college: Extending possibilities for all students.* Albany: State University of New York Press.

Hall, K. & Bucholtz M. (Eds.). (1995). *Gender articulated: Language and the socially constructed self.* New York: Routledge.

Hansen, H. (2004, March 15). Granholm, Cherry announce commission on higher education and economic growth. Retrieved April 9, 2004 from www.michigan.gov/printerFriendly/0,1687,7-168—88248—,00.html.

Harcleroad, F. F. (1999). The hidden hand: External constituencies and their impact. In P. G. Altbach, R. O. Berdahl, & P. J. Gumport (Eds.). *American higher education in the twenty-first century.* Baltimore: Johns Hopkins University Press.

Hardiman, R. & Jackson, B. (1997). Conceptual foundations for social justice courses. In B. Adams, L. A. Bell & P. Griffin (Ed.). *Teaching for diversity and social justice.* New York: Routledge.

Harding, S. (Ed.). (1987). *Feminism and methodology: Social science issues.* Bloomington: Indiana University Press.

Harre, & Slocum. (2003). Disputes as complex social events. *Common Knowledge, 9*(1), 100–118.

Harro, B. (2000a). The cycle of socialization. In B. Adams, Castañeda, Hackman, Peters, & Zúñiga (Ed.). *Readings for diversity and social justice* (pp. 15–20). New York: Routledge.

Harro, B. (2000b). The cycle of liberation. In B. Adams, Castañeda, Hackman, Peters, & Zúñiga (Ed.). *Readings for diversity and social justice* (pp. 463–469). New York: Routledge.

Hart, C. (1998). *Doing a literature review: Releasing the social science research imagination.* Thousand Oaks: Sage.

Herek, G. M. (2001). Internalized homophobia among gay men, lesbians, and bisexuals. In M. Adams, W. J. Blumenfeld, H. W. Hackman, M. L. Peters, X. Zúñiga (Eds). *Readings for diversity and social justice,* pp. 276–281. New York: Routledge.

hooks, b. (1984/2000). *Feminist theory from margin to center.* Cambridge: South End Press.

———. (2000). *Where we stand: Class matters.* New York: Routledge.

hooks, b. & West, C. (1991). *Breaking bread: Insurgent black intellectual life.* Boston: South End Press.

Hurtado, S. (2007). Linking diversity with the educational and civic missions of higher education. *The Review of Higher Education. 30*(2), pp. 185–196.

Hurtado, S., & Wathington, H. (2001). Reframing access and opportunity: Problematic state and federal higher education policy in the 1990s. In

D. E. Heller (Ed.). *The states and public higher education policy: Affordability, access and accountability.* Baltimore: Johns Hopkins University Press.

Institute for Higher Education Policy (IHEP). (1998). *Reaping the benefits: Defining the public and private value of going to college.* Washington, DC: Author.

————. (2005). *The investment payoff: A 50-state analysis of the public and private benefits of higher education.* Washington, DC: Author.

Irigaray, L. (1974). *Speculum of the other woman.* Ithaca: Cornell University Press.

Jafee, A. B. (2000). The U.S. patent system in transition: Policy innovation and the innovation process. Research Policy, 29 (4–5), 5331–5557.

Johnstone, B. (2002). Discourse analysis. Malden, MA: Blackwell.

Jones, D. (2002). *Policy alert: State shortfalls projected throughout the decade.* Washington, DC: The National Center for Public Policy and Higher Education.

Jones, S., & McEwen, M. K. (2000). A conceptual model of multiple dimensions of identity. *Journal of College Student Development, 41(4), 405–413.*

Jones, S. R., Torres, V. & Arminio, J. (2006). *Negotiating the complexities of qualitative research in higher education: Fundamental elements and issues.* New York: Routledge.

Kerr, C. (1963/2001). *The uses of the university (5th ed.).* Cambridge: Harvard University press.

Kerrigan, S. (2009). College graduates' perspectives on the effect of capstone service-learning courses. In P. A. Pasque, N. Bowman & M. Martinez, (Eds.). *Critical issues in higher education for the public good: Qualitative, quantitative and historical perspectives.* Kennesaw, GA: Kennesaw State University Press.

Kettering Foundation. (2008). *Democratic civic engagement gathering, February 26–27, 2008.* Kettering, Dayton, OH.

Keynes, J. M. (1936). *The general theory of employment, interest and money.* New York: Harcourt & Brace.

Kezar, A. J. (2004). Obtaining integrity? Reviewing and examining the charter between higher education and society. *The Review of Higher Education, 27(4), 429–460.*

————. (2005). Challenges for higher education in serving the public good. In A. J. Kezar, A. C. Chambers, & J. Burkhardt, (Eds.). *Higher education for the public good: Emerging voices from a national movement.* San Francisco: Jossey-Bass.

Kincheloe, J. L., & McLaren, P. (2000). Rethinking critical theory and qualitative research. In N. K. Denzin & Y. S. Lincoln (Eds.). *The Sage handbook of qualitative research* (2nd ed.), (pp. 279—313). Thousand Oaks: Sage.

Kincheloe, J. L. & McLaren, P. (2005). Rethinking critical theory and qualitative research. In N. Denzin & Y. Lincoln (Eds.). *The sage handbook of qualitative research* (3rd ed.), pp. 303–342. Thousand Oaks: Sage.

King, D. (1993). Multiple jeopardy: The context of a black feminist ideology. In A. M. Jaggar & P. S. Rothenberg. *Feminist frameworks* (3rd ed.), p. 220. New York: McGraw-Hill.

Komives, S. (2000). Inhabit the gap. *About Campus.* 5(5). San Francisco: Jossey-Bass. 31–32.

Komives, S. R., Lucas, N., & McMahon, T. R. (1998). *Exploring leadership: For college students who want to make a difference.* San Francisco: Jossey-Bass.

KRC Research and Consulting. (2002, October). *National summit: Higher education's role in serving the public good.* Ann Arbor: Author.

Krefting, L. (1991). Rigor in qualitative research: The assessment of trustworthiness. *The American Journal of Occupational Therapy.* (45). P. 214–222.

Krugman, P. (2009). *The return of depression economics and the crisis of 2008.* New York: W. W. Norton.

Labaree, D. F. (1997). *How to succeed in school without really learning.* New Haven: Yale University Press.

―――. (2007). *Education, markets, and the public good: The selected works of David F. Labaree.* New York: Routledge.

Lagemann, E. C. (1983). *Private power for the public good: A history of the Carnegie foundation for the advancement of teaching.* Middletown: Wesleyan University Press.

Laney, J. T. (1981). The other Adam Smith. *Economic Review—Federal Reserve Bank of Atlanta,* 66(7), 26–30.

Lakoff, G. (2006). *Thinking points: Communicating our American values and vision.* New York: Farrar, Straus, & Giroux.

Lakoff, R. (1973). The logic of politeness, or minding your p's and qu's. In C. Corum, T. C. Smith-Stark, & A. Wiser (Eds.), *Papers from the Ninth Regional Meeting of the Chicago Linguistic Society,* pp. 292–305. Chicago: Chicago Linguistic Society.

Lather, P. (2003). Issues of validity in openly ideological research: Between a rock and a soft place. In Y. S. Lincoln & N. K. Denzin (Eds.), *Turning points in qualitative research: Tying knots in a handkerchief,* pp. 185–215. Walnut Creek, CA: AltaMira Press.

Lawrence, B., Weathersby, G. B., & Patterson, V. W. (1970). *Outputs of higher education: Their identification, measurement, and evaluation.* Boulder: Western Interstate Commission for Higher Education, American Council on Education, & Center for Research and Development in Higher Education at Berkeley.

Lee, C. D., Rosenfeld, E., Mendenhall, R., Rivers, A., & Tynes, B. (2004). Cultural modeling as a frame for narrative analysis. In C. Daiute & C. Lightfoot (Eds.). *Narrative analysis: Studying the development of individuals in society,* pp. 39–62. Thousand Oaks: Sage.

Lee, J. B. & Clery, S. (2004). Key trends in higher education. *American Academic,* 1(1), 21–36.

Levine, L. (1996). *The Opening of the American Mind: Canons, Culture and History.* Boston: Beacon Press.

Lewis, E. (2004). Why history remains a factor in the search for racial equality. In P. Gurin, J. S. Lehman, & E. Lewis (Eds.). *Defending diversity: Affirmative action at the University of Michigan.* Ann Arbor: University of Michigan Press.

Lincoln, Y. S. & Guba, E. G. (2000). Paradigmatic controversies, contradictions, and emerging confluences. In N. K. Denzin & Y. S. Lincoln (Eds.). *Handbook of qualitative research* (2nd ed.), pp. 163–188. Thousand Oaks: Sage.

———. (1985). *Naturalistic inquiry.* Beverly Hills: Sage.

Lorde, A. (1983). There is no hierarchy of oppressions. In J. Andrzejewski (Ed.). *Oppression and social justice: Critical frameworks.* Needham Heights: Ginn Press.

Love, P. G., & Estanek, S. M. (2004). *Rethinking student affairs practice.* San Francisco: Jossey-Bass.

Luft, J. (1970). *Group processes: An introduction to group dynamics* (2nd ed.). Palo Alto: National Press Books.

MacDonald, E. (2002). Gender theory and the academy: An overview. In A. M. Martinez Aleman & K. A. Renn (Eds.). *Women in higher education: An encyclopedia* (pp. 71–77). Santa Barbara: ABC-CLIO.

MacLeod, J. (1987). *Ain't no makin' it: Leveled aspirations in a low-income neighborhood.* Boulder: Westview Press.

Magolda, P., & Abowitz, K. K. (1997). Communities and tribes in residential living. *Teachers College Record, 99,* 266–310.

Markus, G. B., Howard, J. P. F., & King, D. C. (1993). Integrating community service and classroom instruction enhances learning: Results from an experiment. *Educational Evaluation and Policy Analysis, 15*(4), 410–419.

Marshall, C. (1999). Researching the margins: Feminist critical policy analysis. *Educational Policy, 13*(1), 59–76.

Marshall, C., & Rossman, G. B. (1999). *Designing qualitative research, 3rd edition.* Thousand Oaks: Sage.

Marwell, G. & Ames, R. E. (1981). Economists free ride, does anyone else?: Experiments on the provision of public goods, IV. *Journal of Public Economics, 15*(3), 16.

Maslow, A. H. (1943). A theory of human motivation. *Psychological Review.* 50. 370–396.

Maxwell, K. E., Traxler-Ballew, A., & Dimpoulos, F. (2004). Intergroup dialogue and the Michigan community scholars program: A partnership for meaningful engagement. In J. Galura, Pasque, P. A., Schoem, D., & Howard, J. (Ed.). *Engaging the whole of service-learning, diversity and learning communities.* Ann Arbor: OCSL Press.

McLaren, P. (1993). *Schooling as a ritual performance: Towards a political economy of educational symbols and gestures* (2nd ed.). New York: Routledge.

McLaren, P. (1994). *Life in schools: An introduction to critical pedagogy in the foundations of education* (2nd ed.). White Plains, NY: Longman.

McLaughlin, D. & Tierney, W. G. (Ed.). (1993). *Naming silenced lives: Personal narratives and the process of educational change.* New York: Routledge.

McMahon, W. W. (2009). *Higher learning, greater good: The private and social benefits of higher education.* Baltimore: Johns Hopkins University Press.

Melissas, N. (2005). Herd behaviour as an incentive scheme. *Economic Theory, 26*(3), 517.

Miller, G. & Fox, K. J. (2004). Building bridges: The possibility of analytic dialogue between ethnography, conversation analysis and Foucault. In D. Silverman (Ed.). *Qualitative research: Theory, method and practice* (2nd ed.). Thousand Oaks: Sage.

Mitchell, J. C. (1984). Typicality and the case study. In R. F. Ellen (Ed.). *Ethnographic research: A guide to general conduct* (pp. 238–241). New York: Academic Press.

Monaci, M., Magatti, M., & Caselli, M. (2003). Network, exposure and rhetoric: Italian occupational fields and heterogeneity in constructing the globalized self. *Global Networks, 3*(4).

Milner, H. R. (2007). Race, culture, and researcher positionality: Working through dangers seen, unseen, and unforeseen. *Educational Researcher, 36*(7), 388–400.

Morris, M., (2000). Dante's left foot kicks Queer Theory into gear. In S. Talburt & S. Steinberg (Eds.). *Thinking Queer, pp.* 15–32. New York: Peter Lang.

Mortenson, T. (2006, March 29). Retrieved on April 6, 2006 on http://postsecondaryopportunity.blogspot.com

Myers, L., Speight, S., Highlen, P., Cox, C., Reynolds, A., Adams, E., & Hanley, T. (1991). Identity development and world view. *Journal of Counseling and Development, 70,* 54–63.

Nadeau, J. (1998). *Families making sense of death.* Thousand Oaks: Sage.

National Association for Equal Opportunity in Higher Education. (April 2009). *35th national conference on Blacks in higher education.* Washington, DC.

National Forum on Higher Education for the Public Good. (2002). *National Leadership Dialogue Series.* Retrieved on March 8, 2006 from http://www.thenationalforum.org/projects_nlds.shtml

Nicholson, M. & Pasque, P. A. (2010). Feminist perspectives in higher education. In P. A. Pasque & M. E. Nicholson, *Empowering women in higher education and student affairs: Theory, research, narratives and practice from feminist perspectives.* Sterling, VA: Stylus.

Nussbaum, M. C. (1999). *Sex & social justice.* New York: Oxford University Press.

Obama, B. H. (2009, July 14). *Remarks from the president on the American graduation initiative.* Washington, DC: The White House, Office of the Presidential Secretary. Retrieved on July 19, 2009 from http://www.whitehouse.gov/the_press_office/Remarks-by-the-President-on-the-American-Graduation-Initiative-in-Warren-MI/

Owen, D. S. (2009). Privileged social identities and diversity leadership in higher education. *The Review of Higher Education. 32*(2). 185–207.

Padilla, R. V. (2008). *Res publica*: Chicano evolving poetics of the public good. In K. P. González & R. V. Padilla, (Eds.). *Doing the public good: Latina/o scholars engage in civic participation,* pp. 13–24. Sterling, VA: Stylus.

Parker, L. (1998). Race is...race ain't: An exploration of the utility of critical race theory in qualitative research in education. *Qualitative Studies in Education.* 11(1), 43–55.

Parker, W. (2003). *Teaching democracy: Unity and diversity in public life.* New York: Teachers College Press.

Pasque, P. A. (2010). Collaborative approaches to community change. In H. E. Fitzgerald, D. L. Zimmerman, C. Burack, & S. Siefer (Eds.). *Handbook of engaged scholarship: Contemporary landscapes, future directions: Volume II: Community-campus partnerships.* Lansing: Michigan State University Press.

———. (2009, May). Voices from women leaders: Multiple constructions of social identity and leadership through grounded theory and performance ethnography. Paper presented at the *International Congress of Qualitative Inquiry,* Urbana-Champagne, IL.

———. (2007). Seeing the educational inequities around us: Visions toward strengthening the relationships between higher education and society. In St. John, E. P. (Ed.). *Readings on Equal Education* (Vol. 22), pp. 37–84. New York: AMS Press.

———. (2005, November). A typology and critical analysis of conceptualizations of higher education for the public good. Paper presented at the *Association for the Study of Higher Education,* Philadelphia, PA.

Pasque, P. A. & Nicholson, M. (Eds.). (2010). *Empowering Women in higher education and student affairs: Theory, research, narratives and practice from feminist perspectives.* Sterling, VA: Stylus.

Pasque, P. A. & Rex, L. A. (2010). Complicating "just do it": Leader's frameworks for analyzing higher education for the public good. *Higher Education in Review, 7,* 47–79.

Pathways to College Network. (2004). *A shared agenda: A leadership challenge to improve college access and success.* Washington, DC: Author.

Paulsen, M. B., & Toutkoushian, R. K. (2007). Overview of economic concepts, models, and methods for institutional research. In R. K. Toutkoushian & M. B. Paulsen (Eds.). *Applying economics to institutional research. New Directions for Institutional Research* 132. 5–24.

Peräkylä, A. (2005). Analyzing talk and text. In N. K. Denzin & Y. Lincoln (Eds.). *The SAGE handbook of qualitative research* (pp. 869–886). Thousand Oaks: Sage.

Perry, J. L., & Katula, M. C. (2001). Does service affect citizenship? *Administration & Society, 33*(3), 330–365.

Pincus, F. L. (2000). Discrimination comes in many forms: Individual, institutional and structural. In Adams, Blumefield, Castañeda, Hackman, Peters & Zúñiga (Eds.). *Readings for diversity and social justice.* New York: Routledge.

Pitkin, H. F. & Shumer, S. M. (1982). On participation. *Democracy*, 2, 43–54.

Porter, K. (2002). The value of a college degree. *ERIC Digest*.

Powell, W. W., & Clemens, E. S. (1998). *Private action and the public good*. New Haven: Yale University Press.

Public Sector Consultants, Inc. (2003). *Michigan's higher education system: A guide for state policy makers*. Lansing: Author.

Putnam, R. D. (1995). Bowling alone: America's declining social capital. *Journal of Democracy*, 6(1), 65–77.

———. (2001). *Bowling alone: The collapse and revival of American community*. New York: Simon & Schuster.

Ramaley, J. A. (2006). Governance in a time of transition. In W. G. Tierney (Ed.). *Governance and the public good*, pp. 157–178. Albany: State University of New York Press.

———. (2005). Scholarship for the public good: Living in Pasteur's quadrant. In A. J. Kezar, A. C. Chambers, & J. Burkhardt (Eds.). *Higher education for the public good: Emerging voices from a national movement*. San Francisco: Jossey-Bass.

Ramazanoğlu, C. with Holland, J. (2002). *Feminist methodology: Challenges and choices*. Thousand Oaks: Sage.

Reason, R., Broido, E., Davis, T., & Evans, N. (Eds.). (2005). *Developing social justice allies*. San Francisco: Jossey-Bass.

Rex, L. A., Steadman, S., & Graciano, M. K. (2006). Researching the complexity of classroom interaction. In J. Green, G. Camilli, & P. Elmore (Eds.). *Handbook of complementary methods for research in education*. Washington, DC: American Educational Research Association.

Rhoades, G. & Slaughter, S. (2004). Academic capitalism in the new economy: Challenges and choices. *American Academic*, 1(1), 37–59.

Richardson, L., & St. Pierre, E. A. (2005). Writing: A method of inquiry. In N. K. Denzin & Y. S. Lincoln (Eds.). *The sage handbook of qualitative inquiry* (3rd ed.), pp. 959–978. Thousand Oaks: Sage.

Ropers-Huilman, R. (2002). Feminism in the academy: Overview. In A. M. M. Alemán & K. A. Renn (Eds.). *Women in higher education: An encyclopedia* (pp. 109–118). Santa Barbara: ABC-CLIO.

Rosenstone, S. J. (2003). The idea of a university. In D. R. Lewis & J. Hearn (Eds.). *The public research university: Serving the public good in new times*. New York: University Press of America.

Rossman, G. B., & Rallis, S. F. (2003). *Learning in the field: An introduction to qualitative research* (2nd ed.). Thousand Oaks: Sage.

Roulston, K. J. (2004). Ethnomethodological and conversation analytic studies. In K. deMarrais & S. D. Lapan (Eds.). *Foundations for research: Methods of inquiry in education and the social sciences*, pp. 139–160. New Jersey: Lawrence Erlbaum Associations.

Rowley, L. L. (2000). The relationship between universities and black urban communities: The class of two cultures. *Urban Review*, 32(1), 45–62.

Rowley, L. L., & Hurtado, S. (2003). Non-monetary benefits of undergraduate education. In D. R. Lewis & J. Hearn (Ed.). *The public research university: Serving the public good in new times.* New York: University Press of America.

Rudolph, F. (1962/1990). *The American college and university: A history.* Athens: University of Georgia Press.

Russi, G. (2004). *2004 state of the public universities address,* Detroit, Michigan.

Salzman, P. C. (2002). On reflexivity. *American Anthropologist, 104*(3), 805–813.

Sarbin, T. R. (2004). The role of imagination in narrative construction. In C. Daiute & C. Lightfoot (Eds.). *Narrative analysis: Studying the development of individuals in society,* pp. 5–20. Thousand Oaks: Sage.

Saxonhouse, A. W. (1992). *Fear of diversity: The birth of political science in ancient Greek thought.* Chicago: University of Chicago Press.

Schlesinger, A. M. (1998). *The disuniting of America: Reflections on a multicultural society.* New York: W. W. Norton.

Schoem, D., Hurtado, S., Sevig, T., Chesler, M., & Sumida, S. H. (2001). Intergroup dialogue: Democracy at work in theory and practice. In D. Schoem, & S. Hurtado (Eds.). *Intergroup dialogue: Deliberative democracy in school, college, community, and workplace,* pp. 1–21. Ann Arbor: University of Michigan Press.

Sevig, T., Highlen, P., & Adams, E. (2000). Development and validation of the self-identity inventory (SII): A multicultural identity development instrument. *Cultural Diversity and Ethnicity Minority Psychology, 6*(2), 168–182.

Sheldon, A. (1993). Pickle fights: Gendered talk in preschool disputes. In D. Tannen (Ed.). *Gender and conversational interaction* (pp. 83–109). New York: Oxford University Press.

Sidanius, J. Levin, S., van Laar, C., & Sears, D. O. (2008). *The diversity challenge: Social identity and intergroup relations on the college campus.* New York: Sage.

Slaughter, S., & Leslie, L. (1997). *Academic capitalism: Politics, policies and the entrepreneurial university.* Baltimore: Johns Hopkins University Press.

Slaughter S. & Rhoades, G. (1996). The emergence of a competitiveness research and development policy coalition and the commercialization of academic science and technology. *Science, Technology and Human Values,* 21(3), 303–339.

———. (2004). *Academic capitalism and the new economy: Markets, state and higher education.* Baltimore: Johns Hopkins University Press.

Smagorinsky, P. (2001). If meaning is constructed, what is it made from? Toward a cultural theory of reading. *Review of Educational Research,* 71(1), 133–169.

Small Business Association. (2004). *21st century jobs.* Retrieved November 27, 2004 from http://www.sba.gov/

Smiley, T. with Robinson, S. (2009). *Accountable: Making America as good as its promise.* New York: Atria.

Smith, A. (1776/1900). *An inquiry into the nature and causes of the wealth of nations.* London: Routledge.

Smith, W. A. (2004). Black faculty coping with racial battle fatigue: The campus racial climate in a post-civil rights era. In D. Cleveland (Ed.). *A long way to go: Conversations about race by African American faculty and graduate students* (Vol. 14, pp. 171–190). New York: Peter Lang.

Solórzano, D. & Bernal, D. D. (2001). Examining transformational resistance through a critical race and latcrit theory framework: Chicana and Chicano students in an urban context. *Urban Education, 36*(3), 308–342.

Soros, G. (2001, October 29). The free market for hope; how a fabled capitalist would take on the root of terror: Poverty. *Newsweek (International ed.),* 56.

Southern Education Foundation. (1995). *Redeeming the American Promise: Report of the Panel on educational opportunity and postsecondary desegregation.* Atlanta: Author

Sowell, T. (2002). *A conflict of visions: Ideological origins of political struggles.* New York: Basic Books.

Stanley, C. A. (2006). Coloring the academic landscape: Faculty of color breaking the silence in predominantly white colleges and universities. *American Educational Research Journal, 43*(4), 701–736.

State Higher Education Executive Officers [SHEEO]. (August 2009). *Higher education policy conference.* Denver, CO.

St. John, E. P. (2003). *Refinancing the college dream: Access, equal opportunity, and justice for taxpayers.* Baltimore: Johns Hopkins University Press.

———. (2006). *Education and the public interest: School reform, public finance, and access to higher education.* Dordrecht, The Netherlands: Springer.

———. (2007). Confronting educational inequality: Lessons learned. In St. John, E. P. (Ed.). *Readings on Equal Education* (Vol. 22). New York: AMS Press.

St. John, E. P. & Hu, S. (2006). The impact of guarantees of financial aid on college enrollment: An evaluation of the Washington State Achievers Program. In E. P. St. John (Ed.). *Readings on equal education: Vol. 21. Public policy and equal educational opportunity: School reforms, postsecondary encouragement, and state policies on postsecondary education,* pp. 211–256. New York: AMS Press.

St. John, E. P., Kline, K. A., & Asker, E. H. (2001). The call for public accountability: Rethinking the linkages to student outcomes. In D. E. Heller (Ed.). *The states and public higher education policy: Affordability, access and accountability.* Baltimore: Johns Hopkins University Press.

Strauss, A., & Corbin, J. (1999). *Basics of qualitative research: Techniques and procedures for developing grounded theory* (3rd ed.). Thousand Oaks: Sage.

Survey: Profit and the public good. (2005). *The Economist, 374*(8410), 15.

Swanson, D. M. (2006). *"Moments of articulation": Taking a look at the contributions of narrative to a critical focus in curriculum.* Paper presented at the Congress of Qualitative Inquiry, Urbana-Champagne, IL.

Talbot, M. M. (1998). *Language and gender: An introduction.* Malden, MA: Blackwell.

Tannen, D. (1993). The relativity of linguistic strategies. In D. Tannen (Ed.). *Gender and conversational interaction.* New York: Oxford University Press.

———. (1994). *Talking from 9 to 5: Women and men in the workplace: Language, sec and power.* New York: Avon.

Tatum, B. D. (2001). The complexity of identity: Who am I? In M. Adams, W. Blumenfeld, R. Castañeda, H. W. Hackman, M. L. Peters, X. Zúñiga. (Eds.). *Readings for diversity and social justice,* pp. 5–9. New York: Routledge.

Taylor, M. C., Trachtenberg, S. J., Weston, L. P., Vedder, R. (June 30, 2009). What is a masters degree worth? *New York Times.* Retrieved August 1, 2009 on http://roomfordebate.blogs.nytimes.com/2009/06/30/what-is-a-masters-degree-worth/?emc=eta1

Taylor, S. (2001). Locating and conducting discourse analytic research. In T. Wetherell, & Yates (Eds.). *Discourse as data: A guide for analysis.* Thousand Oaks: Sage.

Thomas, N. L. (2004). Boundary-crossers and innovative leadership in higher education. In J. Galural, P. A. Pasque, D. Schoem, & J. Howard (Eds.). *Engaging the whole of service-learning, diversity, and learning communities.* Ann Arbor: OCSL Press.

Tierney, W. G. (2006a). The examined university: Process and change in higher education W. G. Tierney (Ed.). *Governance and the public good.* pp. 1–10. Albany: State University of New York Press.

———. (2006b). Trust and academic governance: A conceptual framework. In W. G. Tierney (Ed.). *Governance and the public good.* pp. 179–198. Albany: State University of New York Press.

———. (2003). Remembrance of things past: Trust and the obligations of the intellectual. *The Review of Higher Education, 27*(1), 1–15.

———. (1994). On method and hope. In A. Gitlin (Ed.). *Power and method: Political activism and educational research,* pp. 97–115. New York: Routledge.

Tong, R. (2009). *Feminist thought: A more comprehensive introduction.* (3rd ed.). Boulder: Westview Press.

Torres, C. A. (1998). *Democracy, education, and multiculturalism.* New York: Rowman & Littlefield.

Toutkoushian, R. K. & Shafiq, M. N. (2007). *Using economics to inform the public agenda on the allocation of state funding for higher education.* A paper presented at the annual meeting of the Association for the Study of Higher Education. Louisville, KY.

Trenholm, S. & Jensen, A. (1992). *Interpersonal communication* (2nd ed.). Belmont: Wadsworth.

United Nations Refugee Agency. (2009). *Governance and organization.* Retrieved on June 29, 2009 from www.unhcr.org.

U.S. Department of Education. (2006). *A test of leadership: Charting the future of U.S. higher education. A report of the commission appointed by Secretary of Education Margaret Spellings.* Washington, DC: Author.

Urrieta, L. (2008). Agency and the game of change: Contradictions, *consciencia,* and self-reflection. In K. P. González & R. V. Padilla, eds. *Doing the public good: Latina/o scholars engage in civic participation.* pp. 83–96. Sterling, VA: Stylus.

Valdes, F., Culp, J. M., & Harris, A. P. (Eds.). (2002). *Crossroads, directions, and a new critical race theory.* Philadelphia: Temple University Press.

Vygotsky, L. S. (1978). *Mind in society: The development of higher psychological processes.* Cambridge: Harvard University Press.

W. K. Kellogg Foundation. (2002). *Leadership for civil society: A first in a series of dialogues.* Battle Creek: Author.

Wackwitz, L. A., & Rakow, L. F. (2004). Feminist communication theory: An introduction. In L. F. Rakow & L. A. Wackwitz (Eds.). *Feminist communication theory: Selections in context* (pp. 1–10). Thousand Oaks: Sage.

Wagner, P. (2004). Higher education in an era of globalization: What is at stake? In F. K. Odlin, & P. T. Manicas (Eds.). *Globalization and higher education.* Honolulu: University of Hawai'i Press.

Walker, A. & Parmar, P. (1993a). Warrior marks: Female genital mutilation and the sexual blinding of women. San Diego: Harvest Books.

Walker, A. (Producer) & Parmar, P. (Director). (1993b). *Warrior marks.* New York: Women Make Movies.

Walton, M. D., & Brewer, C. L. (2001). The role of personal narrative in bringing children into the moral discourse of their culture. *Narrative Inquiry, 11*(2), 307–334.

Walton, M. D., Weatherall, A., & Jackson, S. (2002). Romance and friendship in pre-teen stories about conflicts: "We decided that boys are not worth it." *Discourse & Society, 13*(5).

Ward, D. (2006, August 10). *Statement by American Council on Education president David Ward on the final meeting of the Spellings Commission on the future of higher education* [Press release]. Retrieved July 15, 2009, from http://www.acenet.edu/AM/Template.cfm?Section=Press_Releas es2&CONTENTID=17767&TEMPLATE=/CM/ContentDisplay.cfm

Weerts, D. J., & Sandmann, L. R. (2008). Building a two-way street: Challenges and opportunities for community engagement at research universities. *The Review of Higher Education, 32*(1), 73–106.

Weiss, A. (1995). Human capital vs. signaling explanations of wages. *The Journal of Economic Perspectives, 9*(4), 133–154.

Weiss, J. D. (2004). *Public schools and economic development: What the research shows.* Cincinnati: KnowledgeWorks Foundation.

Weissbourd, R., & Berry, C. (2004a). *The changing dynamics of urban America.* Chicago: CEOs for Cities an Alliance for a New Urban Agenda.

Weissbourd, (2004b). *Grads and fads: The dynamics of human capital location*. Washington, DC: Brookings Institute Press.

White, J. S. (2005). Pipeline to pathways: New directions for improving the status of women on campus. *Association of American Colleges and Universities* (Winter).

Wiest, L. R., Abernathy, T. V., Obenchain, K. M., & Major, E. M. (2006). Researcher study thyself: AERA participants' speaking times and turns by gender. *Equity and Excellence in Education, 39*(4), 313–323.

Wildman, S. M., & Davis, A. D. (2000). Language and silence: Making systems of privilege visible. In M. Adams, W. J. Bluenfield, R. Castañeda, H. W. Hackman, P. M. L. & Z. Zúñiga (Eds.). *Readings for diversity and social justice: An anthology on racism, antisemitism, sexism, heterosexism, ableism, and classism* (pp. 50–60). New York: Routledge.

Yosso, T. J. (2005). Whose culture has capital?: A critical race theory discussion of community cultural wealth. *Race, Ethnicity, and Education, 8*(1), 69–91.

Zemsky, R. M. (2005). The dog that doesn't bark: Why markets neither limit prices nor promote educational quality. In J. C. Burke (Ed.). *Achieving accountability in higher education*: Balancing public, academic, and market demands, pp. 275–295. San Francisco: Jossey-Bass.

INDEX